CLIMATE CHANGE
Changing Dimensions of Law and Policy

CLIMATE CHANGE

CHANGING DIMENSIONS OF LAW AND POLICY

Stellina Jolly
Amit Jain

Prints Publications Pvt Ltd
New Delhi

Published by

Prints Publications Pvt Ltd
Viraj Tower-2, 4259/3, Ansari Road,
Darya Ganj, New Delhi-110002
Tel.: +91-11-45355555
Fax: +91-11-23275542
E-mail: contact@printspublications.com
Website: www.printspublications.com

First Edition : 2022 (Hardbound)

ISBN: 978-93-936740-6-7

Price: ₹ 1295/-

Published and Printed by Mr. Pranav Gupta (Director) on behalf of Prints Publications Pvt Ltd, New Delhi.

PREFACE

Climate change, contributed mainly by anthropogenic causes has been recognized and acknowledged as one of the biggest threat for humanity. In fact the adverse affects of climate change are already been felt in various parts of the world affecting people and the earth irreversibly. The unfortunate aspect is that scientific evidences and predictions suggest that climate change will only get worser in the future. The adverse consequences of climate change will not be limited to some portion of earth it will affect both big and small, rich and poor nations.

In legal and policy response to climate change progress has been made. Several International Agreements and Protocols, Regional and National Policy initiatives, Change in Business Practices. An Increased Public Awareness brought about by organizations, NGOs and communities. In their response to combat climate change several theories and principles ranging from human right doctrines, constitutional law, and environmental law have been advocated. As a result a new Climate change law is emerging.

Climate change law and policy is a new and rapidly developing area of law. The law of climate change is being constructed at the intersection of several subjects including science and humanities, and various areas of law, including environmental law, energy law. business law, and international law. Any effort to address climate change has to touch upon these various areas and its relationship with one another. This book ideals with some of these basic elements of climate change law.

The present book is an attempt to study the various policy and legal responses to climate change and related issues arising out of it. The book brings out some of the facets of climate change and its possible effect on humanity. The book also brings out salient features of International and Indian legislative responses. It also reveals various short comings and grave legal vacuum prevailing in this area. The book brings into sharp focus the need for bringing about comprehensive measures steps to mitigate and combat the affects of climate change.

The present book is designed to meet the requirements of Scientific and Legal Professionals, Research Scholars, Post Graduate Students, Law Students, Judicial Officers etc.

The authors express a deep sense of gratitude to friends, colleagues, and parents for their constant encouragement and valuable suggestions. We are very grateful to Mr. Pranav Gupta, Director of Prints Publication Pvt. Ltd. for his kind cooperation and support and for getting the book through the press most expeditiously.

Stellina Jolly

perfectstellina@yahoo.co.in

Amit Jain

Contents

INTRODUCTION

Climate change emerged on the political agenda in the mid-1980s with the increasing scientific evidence of human interference in the global climate system and with growing public concern about the environment. The United Nations Environment Programme (UNEP) and the World Meteorological Organizations (WMO) established the Intergovernmental Panel on Climate Change (IPCC) to provide policy makers with authoritative scientific information in 1988. The UN general assembly responded to there warnings by launching series of negotiations to formulate an International treaty on global climate protection, which finally resulted in completion of the United Nations Framework Convention on Climate Change (UNFCCC) in May 1992 which proposed strategies and laid the framework for the ultimate stabilization of greenhouse gases in the atmosphere, while recognizing the common but differentiated responsibilities and respective capabilities, and social and economic conditions of different jurisdictions. The Convention came into force in 1994. Negotiations held under the convention culminated in the subsequent 1997 Kyoto protocol, which came into force in 2005. This protocol emphatically reasserted the importance of stabilizing greenhouse gas concentrations in the atmosphere and laid down concrete reduction targets of green house gases for developed nations while adhering to sustainable development principles.

India is also seriously concerned about climate change since this phenomenon will have substantial adverse impacts on them. Not all possible consequences of climate change are yet fully understood, but the three main categories of impacts will be on agriculture, sea level rise leading to submergence of coastal areas, as well as increased frequency of extreme events. Each of these pose serious threats to India. In a study done by, Kumar and Parikh who examined the impact of climate change on agricultural crop yields, GDP and welfare. They estimated that: without considering the carbon dioxide fertilization effects in yield losses for rice and wheat vary between 32 and 40%, and 41 and 52%, respectively, Large-scale mass departure from coastal zones is anticipated due to the expected submergence of coast-lines as a result of rising sea levels. This will produce in terms of legal language large mass of environmental refugees especially from low-lying regions. Similarly, intrusion of sea-water in the ground water can affect the agricultural production drastically. Parts of India dependent on coastal fishery and agriculture are likely to be adversely affected. Large areas in Mumbai and West Bengal are clearly threatened by submergence of coastal areas.

Climate change is a multifaceted problem which required an equal multi level response. An analysis of International and National policy measures shows that the climate change response was not merely environmental but economic, social, scientific and technical as well. Indeed these responses have played a positive force in mitigation of climate change and in increasing the awareness of people to this future catastrophe. But these policies suffer from serious shortcomings and controversies. One of the major areas of contention in climate discourse has been the divide between developed and developing countries.

Many especially the developing world sees in Kyoto and associated policies, such as carbon trading programmes, Technology transfer opportunities to make money. They further point out that Climate change has been the creation of blind development. For instance they take the case of renewable energy resource a focal point in climate change actions. Developing countries point out that all these years west which was oblivious to any environmental crisis followed a consumption pattern which cared only for development. The fundamental point is that energy is an essential 'factor of production' that is to say it is an important ingredient in all economic activity. Society can either obtain its energy from low-cost sources, such as coal, oil, gas, and to an extent nuclear and hydro, or it can get its energy from high-cost sources, such as wind turbines, solar cells, and biomass fuels like wood, dung, corn-alcohol, etc which is environment friendly and sustainable. If society obtains its energy from low-cost sources, then more resources are available to be spent on other inputs to production including wages which means that economic growth occurs at a faster pace and people earn more money while doing more furfilling jobs. If society obtains its energy from high-cost sources, then there will be correspondingly fewer resources available for wages and growth-enhancing activities. So, by increasing the cost of all forms of energy, many fear Kyoto and similar policies will reduce the total number of jobs, reduce the average level of skill of workers, and reduce the rate of economic growth. They strongly believe that this is hardly a recipe for a more sustainable economy or society.

This book tries to analyse some of these contentious issues, focus of the book ranges from Renewable energy issues, clean development mechanisms to green consumerism, and India's response to Climate Change etc. The basic arrangement of the book will be as under

Chapters 1 : Climate Change and International Legal Frame Work : Issues and Perspectives

Climate Change is one of the most important global environmental challenges, with implications for food production, water supply, health, energy, etc. Human activities, particularly the burning of fossil fuels and clearing of forests, have increased the level of greenhouse gases, the primary contributors to global warming in the atmosphere. This accumulation is changing the Earth's weather patterns, resulting in higher global temperatures, rising sea levels and a potential shift in the distribution of the world's ecosystems. There is a growing need to develop strategies that will reduce current levels of greenhouse gases in the atmosphere and curtail future emissions. The United Nations Framework Convention for Climate Change (UNFCCC) and the Kyoto Protocol represents an International strategy to combat these effects. These laws have played a positive role in mitigating climate change however these measures suffer from certain drawbacks in respect of climate change mitigation and it can be argued that this partly explains the persistence of the problem. This chapter brings out the structure of international climate change regime, including 1992 UN Framework Convention on Climate Change, 1997 Kyoto Protocol and other additional protocols, the guidelines of the Intergovernmental Panel on Climate Change, Global Environmental Facility and procedures of these institutions. The purpose of this chapter is to examine the efficacy of the legal frame work for combating climate change. The chapter explores the strengths and weaknesses of the international laws on climate change and makes appropriate recommendations for its improvement.

Chapter 2 : National Action Plan on Climate Change: India's Legal Response to Global Warming

India has always been in the forefront of climate change discourse internationally and at the domestic level Number of technological and policy measures were initiated aimed at reducing the green house gas emissions which is responsible for climate change. But we had been shying away from drafting any direct policy framework on climate change. India has been under tremendous pressure from several quarters over its climate change policy. The recent national climate change action plan released by the government of India tried to overcome these criticisms. The NAPCC has laid down a national strategy on climate change with regard to adaptation and mitigation of greenhouse gas emissions, both domestically and in the context of global arrangements under the United Nations Framework Convention on Climate Change (UNFCCC). Sustainable development is the key theme running through the breadth of the plan. The plan in a way marks a new direction in India's climate policy. It brings about a more coherent and systematic approach to climate changes and thereby aims to achieve sustainable development. The plan will definitely give a boost to India's suitability at the global level to augment our bargaining control with the developed nations, especially for providing clean technologies at cheaper rates. At the same time certain serious shortcomings in the plan raises various questions marks over the efficacy of the plan. The chapter critically analyse the NAPCC and see how far this plan will be instrumental in bringing sustainable development without compromising our development needs.

Chapters 3 : Changing Dimensions of Renewable Energy Policy in Response to Climate Change

Fossil fuel use is the largest contributor to greenhouse gas (GHG) emissions, the primary cause of global climate change. India's energy system is dominated by fossil fuel with a heavy dependence on Coal. We started laying emphasis on renewable energy in 1970's primarily as a response to the professed rural energy crisis. The renewable energy programme got momentum with the economic liberalization programme which began in the early 1990's with the emphasis shifting from purely subsidy-driven dissemination programme to technology promotion through the commercial route. In fact ours is the only country with a specific ministry for renewable energy. Despite various policy and technological initiatives including various Electricity Acts, Tariff policies, Tax incentives the renewable energy sector is witnessing slow growth. Key reasons include multiplicity of agencies, skewed incentive structure, and poor implementation capability. The newly launched National Action plan on climate change also emphasis on renewable energy. The chapter brings out the critical relationship between renewable energy and climate change and analyse the changing dimensions of renewable energy policies in India. The chapter evaluates the various impediments in the implementation of these policies. Finally chapter comes out with certain suggestions and recommendations which will help in achieving 100,000 MW by 2050 as proposed in the National Action Plan on Climate Change.

Chapters 4 : Solar Power Policy in India: A Scientific, Environmental and Legal Analysis

In most parts of India, clear sunny weather is experienced 250 to 300 days a year. The annual global radiation varies from 1600 to 2200 kWh/m^2 which is comparable with radiation received in the tropical and sub-tropical regions. The equivalent energy potential is about 6,000 million GWh of energy per year. In spite of having great potential for solar energy production, the progress registered in this regard is very miniscule. The Ministry of New & Renewable Energy (MNRE) has issued guidelines for Generation based incentive for Grid Interactive Solar Photovoltaic power generation projects and Grid Interactive Solar Thermal power generation projects in January 2008 and March 2008 respectively. Different states are also having diverse policies with regard to solar energy. NAPCC identifies Solar as one of the focal mission. Various lacunas plague our solar policies. Many of these policies have been passed without taking into account a detailed feasibility study and technology identification along with proper solar radiation resource assessment. Further current status of international technology and its availability and financial and commercial feasibility in the context of India is not clear. This chapter tries to answer these problems and propose a comprehensive enable policy for SPP based on the guidelines issued by MNRE and NAPCC document.

Chapters 5 : Solid Waste management: A Challenge to Health, Environment and Climate Change

Solid waste problem existed from the historical times, but generally received low priority. The growing urbanization and explosion in world population has changed the quantity and nature of solid waste and management strategy from mainly a low priority, localized issue to an internationally pervasive social problem. It is now accepted that solid waste issue is not a mere problem of pollution, hygiene and health but if adequate steps are not taken for its scientific disposal and management this will adversely contribute to climate change. India currently is facing a municipal solid waste dilemma, for which all elements of the society are responsible. There is an adequate legal framework existing in the country to address MSWM. What is lacking is its implementation which could be the result of a combination of social, technical, institutional and financial issues. This chapter brings out the substantive and procedural framework for solid waste management including judicial response. The purpose of the study is to analyse the efficacy of solid waste management in mitigating climate change. The chapter explores the strengths and weaknesses of the policy measures aimed at mitigating climate change and makes appropriate recommendations for its improvement.

Chapters 6 : CDM : Relevance to Sustainability and Energy Security in Developing Countries

The challenges of how to respond to climate change and ensure sustainable development are currently high on the political agenda among the policy makers. The Clean Development Mechanism (CDM) is part of the global carbon market developing rapidly as part of the Kyoto response towards the mitigation of global warming and to ensure sustainable development among the worlds leading nations. Though hailed as the panacea for global warming, years after the Kyoto Protocol was agreed to, issues related to operationalising CDM are yet to be resolved. This chapter presents the controversies on key CDM issues, namely, those of baselines, additionality, supplementary, application of the sustainable development criteria etc basically from a developing countries perspective. The authors feel that aditionality and baseline clause of CDM undermine the efficient functioning of CDM and disqualify the Developing countries from benefits and creates perverse incentives to introduce new CDM initiatives. After tracing the historical origins and current functioning of CDM Projects under the Kyoto Protocol, the authors observers that current CDM projects is only a half hearted measure to achieve sustainable development and Developing a carbon market. In the context of the CDM, what is required is not just a list of economy-level indicators but a set of micro-level indicators like impacts on local environment, and economy that can be used to measure the extent to which a potential CDM project addresses the objective expressed in Article 12 of the Kyoto Protocol. This would then be formulated into a multi-tier sustainability screening mechanism where projects could be scrutinized on the basis of environmental, social, and economic benefits and access to technology at the international, national, and local levels.

Chapters 7 : Green Consumerism in the Midst of Poverty : Eyewash on Climate Change Mitigation Strategy

Sustainable development, and possibly a reversal of environmental degradation has become the buzzword encircling all political, social, and economical discourse. As an offshoot of it Green consumerism is gaining the momentum Green consumerism can be defined as "the conscious and deliberate choice to make certain consumption pattern taking into account economic consideration. It has been recognized that the pursuit of achieving sustainable and ethical levels of consumption is not only a prudent choice of utilizing accessible reserves of raw materials, but also having substantial correlation within the "carbon sink" capacities of the global environment. Green Consumerism can be looked at from two angles. Firstly at producer's levels who can go green with using recycled materials or contributing money to environment reduction activities and launching green campaigns. Secondly from the consumer's angle opting for environment friendly products. This paper looks at the possibilities and limitations of Green consumerism especially from a third world perspective where acute poverty still persists. The chapter critically analyse the various legislative and policy options initiated to follow eco standards upon companies as well as buyers. Chapter will analyse whether Green consumerism is only eyewash that capitalism has to offer in the face of mounting ecological destruction or motivates manufacturers to be more environmental conscious in their production process.

1

Climate Change and International Legal Framework

Issues and Perspectives

INTRODUCTION

Climate Change is one of the most important global environmental challenges, with implications for food production, water supply, health, energy, etc. Human activities, particularly the burning of fossil fuels and clearing of forests, have increased the level of greenhouse gases, the primary contributors to global warming in the atmosphere. This accumulation is changing the Earth's weather patterns, resulting in higher global temperatures, rising sea levels and a potential shift in the distribution of the world's ecosystems. There is a growing need to develop strategies that will reduce current levels of greenhouse gases in the atmosphere and curtail future emissions. The current international climate change regime comprises a network of agreements and mechanisms. A high water mark in the evolution of this regime was the entry into force in February 2005 of the Kyoto Protocol to the United Nations Framework Convention on Climate Change ("UNFCCC" or "Framework Convention"). However, the Kyoto Protocol will not be the final word on the issue. Its emission limits cover only a fraction of the world's greenhouse gas ("GHG") emissions and those limits expire in 2012. This chapter brings out the structure of international climate change regime, including 1992 UN Framework Convention on Climate Change, 1997 Kyoto Protocol and other additional protocols, the guidelines of the Intergovernmental Panel on Climate Change, Global Environmental Facility and procedures of these institutions. The purpose of this chapter is to examine the efficacy of the legal frame work for combating climate change. The chapter explores the strengths and weaknesses of the international laws on climate change and makes appropriate recommendations for its improvement.

CLIMATE CHANGE: SCIENCE AND CONSEQUENCES

Many scientists believe that part of the earth's atmosphere is acting as an insulation trapping sufficient infrared radiation to keep the global average temperature in a comfortable range. The insulation is the collection of the six atmospheric gases called "greenhouse gases" (GHGs), which consist of carbon dioxide, methane, nitrous oxide, and fluorinated gases (hydrofluorocarbons, perfluorocarbons and sulphur hexafluoride). Rising concentration of GHG emissions has been detected, and serious negative consequences have been predicted. Over the past several decades, increasing Greenhouse gases are produced by a variety of human activities, most importantly the burning of fossil fuels and are called the enhanced greenhouse effect.[1] The greenhouse effect has been known since the turn of the 19th century, when a Swedish chemist named Svante Arrhenius first described it.[2] The enhanced greenhouse effect will increase the average temperature of the

---

1 Albert K Barts, Climate in Crisis: The Greenhouse Effect and What We Can Do, Book Publishing Company, p 13,1990. There are various experts who believe that Green house effect does not exist SEE Julie Wheldon, "Greenhouse Effect is a Myth, say Scientists" available at http://www.dailymail.co.uk/sciencetech/article-440049/Greenhouse-effect-myth-say-scientists.html(Last accessed on 3-04-2009).

2 Sebastian Oberthur & Hermann E. Ott, The Kyoto Protocol: International Climate Policy for the 21st Century 3, Springer (1999).

Earth's surface gradually also known as the global warming effect, and will possibly change the global heating and cooling system and resulting in climate change.[3]

The issue of climate change has now captured the popular imagination like no other cause in living memory and is been characterized as 'the defining human development challenge for the 21st century.' [4] The issue of global climate change came to the fore in the mid-1980s when it was first discussed in the UN General Assembly, since then it has become a contentious political issue, however, the science has become increasingly certain regarding the occurrence of climate change and its adverse impacts.

The world became increasingly conscious of the climate change when IPCC released its first report which noted that green house effects are certain.[5] Intergovernmental Panel on Climate Change ("IPCC"), reports are considered to be the most reliable source of scientific data. The latest Fourth Assessment Report of the Intergovernmental Panel on Climate Change (IPCC), released in 2007, warned that the warming of the climate system is unequivocal and accelerating.[6] The global average temperature has increased by 0.74 centigrade in the last century, the largest and fastest warming trend in the history of the Earth.[7] It also estimates an increase of 1.4 to 5.8 degrees Celsius (2.5 to 10.4 degrees Fahrenheit) over the next century.[8] The IPCC is currently starting to outline its

---

[3] Tietenberg, T. H., "Economic Analysis and Climate Change." Environment and Development Economics 3(3): 402-405,(1998).

[4] Fighting Climate Change: Human Solidarity in a Divided World, UNDP, Human Development Report, 2007-8.

[5] JT Houghton, GJ Jenkins and JJ Ephraums, Scientific Assessment of Climate change - Report of Working Group I, IPCC 1990 Cambridge University press.

[6] Fourth Assessment Report of the Intergovernmental Panel on Climate Change (Cambridge University Press, 2007).

[7] *Id*

[8] *Id*

Fifth Assessment Report (AR5) which will be finalized in 2014.

What are the impact and consequences of this climate change? Experts predicts that as ice caps and glaciers melt, sea levels are expected to rise 9 to 88 centimeters, or 4 to 35 inches.[9] The effects of sea level rises of that magnitude may include widespread coastal flooding, the further melting of glaciers and ice caps, changes in rainfall patterns, more variable and extreme weather, and possibly more catastrophic events--shifts in the Gulf Stream or the collapse of the Greenland or Antarctic ice sheets.[10] Climate change will, among other impacts, increase the severity of droughts, land degradation, desertification, the intensity of floods and tropical cyclones, the incidence of malaria and heat-related mortality, and decrease crop yield and food security. At stake are basic elements of human life concerning access to water, production of food, health matters and habitation issues.[11] There is also increasing certainty that, as the climate system warms, poorer nations, and the poorest within them, will be the worst affected. Furthermore, such changes may have significant impacts on the global economy and international trade in particular. It is crucial to underline here that the effects of climate change will be distributed unevenly around the world. Whereas few developed countries can feel some benefits depending on the scale of global warming, the poorest nations will experience extremely adverse effects almost immediately.[12] Such exposition of unfair vulnerability, especially in the light of unequal contribution to production of greenhouse gas (GHG) emissions, will create more complications in the North-South relations, as well as deepen a division on winners and losers of the climate change and

---

9 *Id*

10 *Supra* note 3 see Titenberg.

11 *Id*

12 *Stern Review on The Economics of Climate Change* Stern Report, part 11 Impact of Climate change,(2007).

international trade. Many effects are already being felt. Scientists have noted earlier flowering of plants and bird egg-laying, thawing of Alaskan tundra, and shrinking of mountain glaciers atop Mt. Kilimanjaro, in the Andes, and in the Rocky Mountains.[13]

THE U.N. FRAMEWORK CONVENTION ON CLIMATE CHANGE: DECLARATION OF PRINCIPLES

The characteristics of climate change create unique legal challenges, and inform what legal responses are appropriate. Addressing the consequences of global warming through climate change management is not a regulatory field that originally belonged to international law - initially there have been various attempts to tackle its symptoms on the domestic level.[14] Later it was felt that climate change is a global problem necessitating a global response. [15] There are basically certain constraints in developing a global consensus on this issue. Firstly, not all countries have an equal interest in reducing emissions, creating difficulties in a consensus-based international legal process. The problem is also long-term, as CO_2 emissions, on average, remain in the atmosphere for about one hundred years [16] and other anthropogenic gases persist for thousands of years. Left unchecked, consequences

[13] UNFCCC Preliminary Report, (A Guide to the Climate Change Convention and its Kyoto Protocol describing present effects that have occurred due to the change in climate), available at http://unfccc.int/resource/guideconvkp_p.pdf

[14] *Supra* note 5 See IPCC Report.

[15] Bert Metz, *Climate Change: Mitigation: Contribution of Working Group III to the third assessment report of the Intergovernmental Panel on Climate Change.* Cambridge, Cambridge University Press,(2001).

[16] The atmospheric lifetime of CO_2 varies due to different removal processes, but the dominant CO_2 decay lifetime is approximately 100 years. Michael Grubb et al., The Kyoto Protocol: A Guide and Assessment 73-74 & n.7 (1999).

such as sea level rise can be irreversible.[17] Furthermore, nearly the full range of human activities are associated with GHG emissions, including transportation, industrial activities, and electric power usage. Thus, lawmaking in this field implicates essential interests such as economic development and national security. Further, uncertainty surrounding climate change is pervasive.[18] Although scientists largely agree on the causes and general effects of climate change, the localized effects of climate change are hard to predict, as are the costs and benefits of reducing GHG emissions.[19] Because the nature and scale of the climate change problem pose such challenges to domestic lawmaking and international treaty making, it is not surprising that the global agreements needed to address climate change are only partially formed.

Gradually members of the international community realized that, due to the trans-boundary nature of climate change and its ensuing consequences as well as the high costs involved in mitigation, [20] climate change can only be effectively addressed by complementing domestic measures through institutionalized forms of transnational collaboration.[21] As a consequence, combating climate change has spilled over from the realm of domestic regulation into the field of international cooperation. The 1979 Geneva Convention on Long-Range Trans-boundary Air Pollution[22] as

---

[17] Supra note 5p 607.

[18] *Id.* at 608-09.

[19] IPCC Working Group III, supra note 23, at 608-09.

[20] M. J. Mace, Chris Hendriks & Roger Coenraads, Regulatory Challenges to the Implementation of Carbon Capture and Geological Storage Eithin the European Union under EU and International Law, 1 International Journal of Greenhouse Gas Control 253 (2007).

[21] Marta D'Auria, Emissions Trading and Polycentric Negotiation, 6 GLOBAL JURIST ADVANCES 1 (2006).

[22] The aim of the Convention is that Parties shall endeavour to limit and, as far as possible, gradually reduce and prevent air pollution including long-range transboundary air pollution. Parties develop policies and strategies to combat the discharge of air pollutants through exchanges of information,

Contd....

well as the 1985 Vienna Convention for the Protection of the Ozone Layer and its 1987 Montreal Protocol on Substances that Deplete the Ozone Layer[23] constituted serious international efforts to face the problem of climate change, adopting novel solutions under international environmental law. However, these conventions targeted only specific aspects of the problem of climate change.

The emerging anxiety for rising CO2 levels in the atmosphere prompted scientists and other representatives to meet at the 1979 First World Climate Conference organized by the World Meteorological Organization (WMO)and Climate change was identified as an "urgent world problem".[24] In 1979 the World Climate Programme was established by the World Meteorological Organization (WMO), the United Nations Environment Programme (UNEP), and the International Council of Scientific Unions.[25] Various scientific and political

Contd. from previous page...

consultation, research and monitoring. Currently, the Convention's priority activities include review and possible revision of its most recent protocols, implementation of the Convention and its protocols across the entire UNECE region.

[23] The Montreal Protocol on Substances that Deplete the Ozone Layer, UNEP programme,The Montreal Protocol on Substances that Deplete the Ozone Layer is an international treaty designed to protect the ozone layer by phasing out the production of a number of substances believed to be responsible for ozone depletion.

[24] The world climate conferences are a series of international meetings, organized by the World Meteorological Organization (WMO), about global climate issues. The first two conferences largely focused on climate change in addition to climate research and forecasting. The next upcoming conference is World Climate Conference-3, to be held in Geneva, Switzerland, 31 August to 04 September 2009. The theme of the upcoming conference is "Climate prediction and information for decision-making". Recognizing the growing evidence for the widespread effects of climate variability and change.

[25] The World Climate Programme (WCP) is an authoritative international scientific programme whose goals are to improve understanding of the climate system and to apply that understanding for the benefit of societies coping with climate variability and change. The World Climate Programme was established following the staging of the First World Climate Conference

Contd....

conferences took place, organized in collaboration with UN Environmental Programme (UNEP) and those meetings helped elevate the problem of climate change from just a scientific concern to a serious international issue. [26] By 1988 there was significant international consensus that climate change must be addressed in a comprehensive framework convention. In 1989, the United Nations issued The Hague Declaration on the Environment, [27] which publicly affirmed the dangers of global atmospheric warming, deterioration of the ozone layer, and climate change. The report called for a solution that was "vital, urgent and global." [28] Most notably, the Declaration foreshadowed the problem of "equity," stating that most emissions originated in industrialized nations where "room for change is the greatest" and "the greatest resources to deal with this problem" exist. [29] and later that year the U.N. General Assembly, in Resolution 43/53, declared that "climate change is a common concern of mankind."[30] As international concern mounted, the United Nations Environment Programme ("UNEP") and World Meteorological Organization ("WMO") established the Intergovernmental Panel on Climate

Contd. from previous page...

in Geneva, Switzerland in February 1979, UNFCCC SECRETARIAT, UNITED NATIONS FRAMEWORK CONVENTION ON CLIMATE CHANGE HANDBOOK 17-19 (2006), available at http://unfccc.int/resource/docs/publications/handbook.pdf(Last accessed on 5-02-2009).

[26] Regina S. Axelrod, David Leonard Downie, and Norman J. Vig, The Global Environment:Institutions, Law, and Policy, 2d ed,(2005).

[27] The United Nations' activities prior to The Hague Declaration added momentum for action. In particular, the United Nations General Assembly adopted the "Protection of the Global Climate" Resolution, catapulting climate change forward as a global issue.

[28] The Hague Declaration on the Environment, March 11, 1989, 28 I.L.M. 1308, 1309.

[29] *Id*

[30] Protection of global climate for present and future generations of mankind, General Assembly, A/RES/43/53,1988 states that climate change is a common concern of mankind, since climate is an essential condition which sustains life on earth.

Change in 1988.[31] The IPCC was established to provide the decision-makers and others interested in climate change with an objective source of information about climate change. Its constituency is made of:

- The governments: The IPCC is open to all member countries of WMO and UNEP. Governments of participate in plenary Sessions of the IPCC where main decisions about the IPCC work programme are taken and reports are accepted, adopted and approved. They also participate in the review of IPCC Reports.
- The scientists: Hundreds of scientists all over the world contribute to the work of the IPCC as authors, contributors and reviewers.
- The people: As United Nations body, the IPCC work aims at the promotion of the United Nations human development goals. [32]

The IPCC has three main groups which each produce a report: Working Group I, which produces the report on the Physical Science Basis of Climate Change; Working Group II, which focuses on Impacts, Adaptation and Vulnerability; and Working Group III, which focuses on Mitigation. [33] In addition, the IPCC produces the Synthesis Report and other Special Reports. Final reports are accepted at a Plenary Session of the IPCC and the Summaries for Policymakers are

---

[31] The IPCC was established to provide the decision-makers and others interested in climate change with an objective source of information about climate change available at *http://www.ipcc.ch/about/index.htm(Last accessed on 3-03-2009).*

[32] *Id*

[33] Intergovernmental Panel on Climate Change, About the IPCC, How the IPCC is Organized, http://www.ipcc.ch/about/how-the-ipcc-is-organized.htm (last accessed on 3-03-2009) see also Intergovernmental Panel on Climate Change, About the IPCC, The IPCC Working Group I, http://www.ipcc.ch/about/working-group1.htm (last accessed on 3-03-2009; Intergovernmental Panel on Climate Change, About the IPCC, The IPCC Working Group II, http:// www.ipcc.ch/about/working-group2.htm (Last accessed on 3-04-2009).

approved line by line. The last steps in the procedure include a consensus by the international community.[34]

The IPCC provides its reports at regular intervals and they immediately become standard works of reference, widely used by policymakers, experts and students. The IPCC First Assessment Report revealed the environmental and socio-economic consequences of climate change, as well as the issue of adaptation to impacts of global warming, and therefore drew attention of the international community which acknowledged the necessity of preventing such changes. [35] The findings of the first IPCC Assessment Report of 1990 played a decisive role leading to the United Nations Framework Convention on Climate Change (UNFCCC).[36]As a result of the IPCC report, in the end of 1990, under resolution 45/212, the UN General Assembly (UNGA) established the Intergovernmental Negotiating Committee for a Framework Convention on Climate Change (INC/FCC). [37]

The process of negotiations under INC/FCCC, took eighteen months and produced a convention document ready for the 1992 Rio Conference, but at the same time this negotiation showed first areas of disputes between bargaining countries. [38] According to Lorraine Elliott, the first area that was considered a question of how to stabilize emissions and reduce their concentration, showed that it was not only a scientific and technical problem but, and most of all, a highly political issue, simply because it brought up revision of economic and particularly energy policies of negotiating

---

[34] Intergovernmental Panel on Climate Change, Principles Governing IPCC Work 2, available at http://www.ipcc.ch/pdf/ipcc-principles/ipcc-principles.pdf.(Last accessed on 3-04-2009).

[35] Lorraine Elliott, *The Global Politics of the Environment, Palgrave Macmillan, New York* 2nd *edt, p 167,(2004)*.

[36] United Nations Framework Convention on Climate Change, 3, May 9, 1992 31 I.L.M. 849, available at http://unfccc.int (last visited 11-02-2009).

[37] "Protection of global climate for present and future generations of mankind" A/RES/45/212,1990.

[38] *Supra* note 35 See Elliott.

countries. The second area of dispute had to confront a problem of responsibility to act and, therefore, it raised questions of who has emitted more GHGs, also in a historical perspective, and who should first start mitigating the problem of climate change. It appeared, that developed countries could not escape the responsibility of their economies and industries that significantly contributed to GHG concentration in the atmosphere. However, the question was whether developing countries, eager to further continue their economic growth, would also considerably contribute to the global pollution and, therefore, should also act sooner or later.[39] The third area of differences cumulated around the issue of scientific uncertainties regarding the global climate change and how they should be included into the negotiation process. After the long drawn negotiations treaty was adopted and opened for signature in the Rio de Janeiro Summit in 1992 and entered into force in 1994. [40] Overall, 186 nations are now parties to the Convention.

The UNFCCC was more a declaration of principles than a detailed and binding treaty reducing emissions.[41] The UNFCCC does not establish binding limits on GHG emissions for any countries. Rather, true to its name, it forms a framework for further action and cooperation on the issue of climate change. The Convention sets the ultimate objective of stabilizing greenhouse gas emissions "at a level that would prevent dangerous anthropogenic interference with the climate system". A key aspect of the treaty is its division of nations into "Annex I" (forty-one industrialized countries and economies in transition, such as the former Soviet Union and Eastern European states) and "Annex II" (twenty- four wealthy members of the Organization of Economic Cooperation and

[39] *Id*

[40] Supra note 36 See UNFCCC.

[41] Sebastian Oberthur & Hermann E. Ott, The Kyoto Protocol: International Climate Policy for the 21st Century 3 (1999).

Development).[42] The remaining countries of the world are described as "non-Annex I" countries. This division became the guiding force even in Kyoto protocol.

The UNFCCC outlines certain commitments only relevant to Annex I Parties. Article 4.2 obliges Annex I Parties to adopt national policies to mitigate climate change and to report on the progress of these policies "with the aim of" returning emissions to their 1990 levels. [43] With a few exceptions, the "base year" for tabulating greenhouse gas emissions has been set as 1990. This became a "soft" commitment; Annex I Parties generally have not met this target, and many missed the target by a wide margin. The UNFCCC requires developing countries to develop in a sustainable manner and address the adverse effects of climate change through adaptation. This responsibility is unique to developing countries in the sense that it requires developing countries to take on board sustainable development.

Article 3 of the Framework Convention sets forth a series of guiding "principles" that attempt to balance the aims of environmental protection, economic development, and the general division of burdens between developed and developing country Parties. One of the principles is the so-called "precautionary principle," which provides that where there are "threats of serious or irreversible damage, lack of full scientific certainty should not be used as a reason for postponing [precautionary] measures."[44] Other principles

[42] Supra note 36 See UNFCCC.

[43] *Id Art 4.2*

[44] *Id* Art 3.3 provides: The Parties should take precautionary measures to anticipate, prevent or minimize the causes of climate change and mitigate its adverse effects. Where there are threats of serious or irreversible damage, lack of full scientific certainty should not be used as a reason for postponing such measures, taking into account that policies and measures to deal with climate change should be cost-effective so as to ensure global benefits at the lowest possible cost. To achieve this, such policies and measures should take into account different socio-economic contexts, be comprehensive, cover all relevant sources, sinks and reservoirs of greenhouse gases and adaptation,

Contd....

relates to intergenerational equity, Public trust, Polluters principle etc.

A fundamental and recurring theme in the UNFCCC is that developed and developing country Parties have "common but differentiated responsibilities and respective capabilities,"[45] reflecting a view that developed countries bear a greater historical responsibility for the accumulation of GHG emissions and have greater capacity to take action. The principle known as "common but differentiated responsibilities" is relatively new to international law. The first use of the phrase in an international law instrument was in 1992 in the United Nations Framework Convention on Climate Change (UNFCCC) and in the Rio Declaration. [46] Five years later, the principle was again used, again in the Kyoto Protocol to the UNFCCC. It addresses the perennial challenge of how nations should share the global commons. It recognizes that there are global resources, such as the atmosphere, which human society has common interest in protecting, but which the obligation to protect will vary in accordance with a nation's level of development, resources, and institutional capabilities. However, with time the meaning attributed to the phrase in the context of the UNFCCC regime changed dramatically. From a concept that started with common responsibilities equitably allocated among nations, in the UNFCCC, the concept shifted in the Kyoto Protocol to

---

Contd. from previous page...

and comprise all economic sectors. Efforts to address climate change may be carried out cooperatively by interested Parties.

[45] UNFCCC, *Id.*, at Art. 3.1. 1provides. The Parties should protect the climate system for the benefit of present and future generations of humankind, on the basis of equity and in accordance with their common but differentiated responsibilities and respective capabilities. Accordingly, the developed country Parties should take the lead in combating climate change and the adverse effects.

[46] United Nations Conference on Environment and Development, Rio Declaration on Environment and Development, Principle 7, June 14, 1992, UN Doc. A/CONF.151/5/Rev.1

operate as a catch phrase justifying imposing no responsibility on developing countries.[47]

The UNFCCC also provides that all Parties will develop and submit national inventories of emissions by sources and removals by sinks,[48] implement national plans that include measures to mitigate climate change,[49] promote and cooperate in technology transfer,[50] and encourage and assist in scientific research on climate change. Countries ratifying the treaty - called "Parties to the Convention" agree to take climate change into account in such matters as agriculture, industry, energy, natural resources, and activities involving seacoasts. The convention has also established an Annual Conference of Parties (COP) to oversee its implementation in the member countries. The Convention recognizes that it is a "framework" document - something to be amended or augmented over time so that efforts to deal with global warming and climate change can be focused and made more effective. The first addition to the treaty, the Kyoto Protocol, was adopted in 1997.[51]

THE KYOTO PROTOCOL: FIRST BINDING MECHANISM

The origins of the Kyoto Protocol can be found in the report of UNFCCC COP-1 in Berlin (1995) which triggered a hot debate that was, tinged by the North-South dimension. The Parties to the Framework Convention determined that a more forceful international response to the threat of climate change was needed. [52] The task of preparing a legal tool for

---

[47] Christopher D. Stone, "Common but Differentiated Responsibilities in International Law", *98 AM. J. INT'L L.* 276, 299-300 (2004).

[48] UNFCCC, *supra* note 36 Art. 4.1(a)

[49] *Id* at Art. 4.1(b).

[50] *Id*, at Art. 4.1(c).

[51] Kyoto Protocol to the United Nations Framework Convention on Climate Change, Feb. 16, 2005, 37 I.L.M. 32, available at http://unfccc.int/resource/docs/convkp/kpeng.pdf (last visited 11-02-2009).

[52] Houghton Climate Change 1995: The Science Of Climate Change: Contribution Of Working Group I To the Second Assessment Of The

Contd....

setting commitments was given to a specially created Ad Hoc Working Group in Berlin mandate (AGBM).[53] One of the most debatable issues of the negotiations under AGBM was a legal and functional definition of GHG emission targets, including what should constitute their basis of reduction and how these emissions should be counted. These negotiations resulted in the Kyoto Protocol, which was adopted by the Parties at UNFCCC COP-3 at Kyoto in 1997. The Protocol outlined emission limits for the Convention's Annex I Parties. However, many key details about the Protocol were not resolved and negotiations continued.

The Kyoto Protocol sets forth binding emission limits for developed country Parties for the period of 2008-2012.[54] Parties effectively have full discretion in developing national measures to meet their limits. Furthermore, they can take advantage of certain "flexible mechanisms," which offer market-based approaches for achieving emission reductions across borders. Each Annex I Party's commitment applies on the basis of a "basket" of six GHGs: carbon dioxide ("CO_2"), methane ("CH_4"), nitrous oxide ("N_2O"), hydrofluorocarbons ("HFCs"), perfluorocarbons ("PFCs"), and sulfur hexafluoride ("SF_6").[55] The Intergovernmental Panel on Climate Change ("IPCC") determined the "global warming potential" of each of these types of GHGs relative to carbon dioxide.[56] The Protocol expresses each Party's limit in the form of a "carbon dioxide equivalent" in tons of GHG emissions. Instead of a single fixed-year limit, the Protocol's emission commitments apply as an annual average to be achieved over a five-year period. This

Contd. from previous page...

Intergovernmental Panel On Climate Change (J.T. Houghton, et al., eds., Cambridge University Press 1995).

53 Supra note 35 See Elliott.

54 Supra note 51 See Kyoto Art 3.

55 Supra note 51 See Kyoto, at Annex A.

56 Intergovernmental Panel on Climate Change, Climate Change 2001: The Scientific Basis 6.1.2 (2001).

approach responds to the concern that a country's GHG emissions could rise or fall in any particular year because of difficult-to-control factors such as the vagaries of the business cycle or weather fluctuations affecting power production.

Article 4 of the Protocol provides that two or more Annex I Parties may agree to fulfill their Article 3 commitments jointly, in which case they become subject to a summed assigned amount, rather than their individual commitments.[57] The European Union ("EU") opted to take advantage of this provision, replacing each Member State's Annex B commitment with a collective commitment known as European Union Bubble mechanism.[58]

Activities that lead to deforestation, or even clearing of agricultural land and disturbance of soils, result in substantial releases of carbon dioxide. For these reasons, issues related to land use, land use change, and forestry ("LULUCF") have been a significant topic of discussion at the COPs. The Protocol embodies a complicated set of compromises on these issues. First, the Protocol provides that, in meeting their Article 3 commitments, Annex I countries only may take into account a finite set of relatively easily-measured activities "net changes in greenhouse gas emissions by sources and removals by sinks resulting from direct human-induced land-use change and forestry activities, limited to afforestation, reforestation and deforestation since 1990." [59]The Protocol did not fully resolve

---

[57] Supra note 51 See Kyoto Art 4.

[58] Sebastian Oberthür, Hermann, The Kyoto Protocol: international climate policy for the 21st century,Springer,141 (1999).

[59] Supra note 51 See Kyoto Article 2.1(a)(ii) requires Annex B Parties to implement policies and measures relating to protection and enhancement of sinks and reservoirs, and promotion of sustainable forest management practices, afforestation, and reforestation, Article 3.3 requires Annex B countries to net sources and sinks associated with reforestation, afforestation, and deforestation activities since 1990 against their non-biotic emissions. See also, Dr. Mark C. Trexler, "Land Use Cange And Forestry Under The Kyoto Protocol: Looking Forward To COP-4", *Linkages Journal Volume* 3 ,No 4,(1998).

whether forest management other than afforestation and reforestation could count toward targeted reductions.

Perhaps the most important international environmental law innovation of the Kyoto Protocol is its establishment and significant reliance on market-based instruments, often referred to as the "flexible mechanisms." These mechanisms are the Article 17 emissions trading system, Article 6 Joint Implementation, and the Article 12 CDM. The rationale for the flexible mechanisms is straightforward. All emissions of GHG have an identical impact on the atmosphere regardless of their location, but the cost of achieving emission reductions varies substantially.

FLEXIBLE MECHANISMS: MARKET BASED GHG REDUCTION PROCESS

One of the flexible mechanisms under the protocol is Emission Trading. Under this each Annex I Party's Article 3 commitment translates into an "assigned amount," of GHG emissions over the five-year commitment period.[60] At the end of the commitment period, an Annex I Party's emissions cannot exceed its assigned amount. The Protocol further provides that a Party's assigned amount can be subdivided into "assigned amount units" ("AAUs"), with each AAU corresponding to the right to emit one carbon dioxide equivalent ton of GHG emissions. Article 17 directs the COP to develop rules under which Annex I can trade AAUs with one another.[61] A concern identified during the Protocol

---

60 *Id,Art 3 provides:* The Parties included in Annex I shall, individually or jointly, ensure that their aggregate anthropogenic carbon dioxide equivalent emissions of the greenhouse gases listed in Annex A do not exceed their assigned amounts, calculated pursuant to their quantified emission limitation and reduction commitments inscribed in Annex B and in accordance with the provisions of this Article, with a view to reducing their overall emissions of such gases by at least 5 per cent below 1990 levels in the commitment period 2008 to 2012.

61 *Id,* Art 17 provides: The Conference of the Parties shall define the relevant principles, modalities, rules and guidelines, in particular for verification,

Contd....

negotiations was that the Article 17 system could create a risk of "overselling." To address this risk, the Protocol provides that an Annex I Party may not engage in trades that would bring its holdings of AAUs or other Kyoto credits below the level of its "commitment period reserve. On the other hand, developing countries were particularly opposed to the concepts of emissions trading, understanding them rather as the rich North's tactic to shirk responsibilities of climate change mitigation efforts. However, it was not the only reason of the South's reluctance and mistrust expressed towards the idea of global emissions trading. First, the Kyoto Protocol, by introducing flexible mechanisms, created a basis for carbon markets and, therefore, blended climate change mitigation efforts with the mainstream neo-liberal economy framing them within the globalization process.[62] Second, for the majority of developing countries, environmental regulations constituted a very new branch of domestic and particularly international law. Moreover, they did not have their own strategies allowing them to tackle various environmental problems. Thus, developing countries did not have time to acquire knowledge and experience regarding new techniques dealing with the climate change.[63] A third concern centers on the nature of its emissions trading elements. The provision in Article 17 for international emissions trading is unlikely to be effective, if indeed it is utilized at all.[64] The entire theory

Contd. from previous page...

reporting and accountability for emissions trading. The Parties included in Annex B may participate in emissions trading for the purposes of fulfilling their commitments under Article 3.

[62] Linnér, B.-O. & Jacob, M., From Stockholmto Kyoto and Beyond: A Review of the Globalisation of Global Warming Policy and North-South Relations. Globalizations 2 (3): 403 -415,(2005).

[63] Farana Yamin, Climate Change and Carbon Markets: A Handbook of Emissions Reduction Mechanisms, Earthscan/James & James (2005).

[64] Hahn, Robert W. and Stavins, Robert N. What has the Kyoto Protocol wrought? The real architecture of international tradable permit markets. Washington, DC: American Enterprise Institute Press, (1999).

behind the claim that a cap-and-trade system is likely to be cost effective depends upon the participants being cost-minimizing entities. In the case of private-sector firms, this is a sensible assumption, because if firms do not seek, and indeed succeed in, minimizing their costs, they will eventually disappear, given the competitive forces of the market. But nation-states can hardly be thought of as simple cost-minimizers—many other objectives affect their decision-making. Furthermore, even if nation-states sought to minimize costs, they do not have sufficient information about marginal abatement costs at the multitude of sources within their borders to carry out cost-effective trades with other countries.[65] In principle, this trading scheme applies to nation states, although the participation of companies is not explicitly excluded. This concept is known as the "carbon market". Smaller "carbon markets" have been established, for example, by the European Union and other groups of countries. These emissions-trading systems are intended to start the process and to link up with the Protocol's global market.

The European Union Emission Trading system is a so-called "cap-and-trade" scheme regulating industrial CO_2 emissions in the EU. The scheme started in January 2005.[66] The first phase runs from 2005-2007, the second one from 2008-2012, coinciding with the first Kyoto Protocol commitment period. About half of EU CO_2 emissions are covered by the scheme.[67] It is targeted at large individual energy-using installations in defined economic sectors: mainly energy production, metals, construction materials, and paper.

---

65 *Id*

66 The European Union Emission Trading System (EU ETS) is the largest multi-national, emissions trading scheme in the world and is a major pillar of EU climate policy. It was established by the Directive 2003/87/EC of the European Parliament and of the Council of 13 October 2003 establishing a scheme for greenhouse gas emission allowance trading within the Community and amending Council Directive 96/61/EC.

67 *Id*

As a market-based mechanism, the ETS ensures that emissions are reduced in the most cost-efficient way possible, the cap giving effective control over total emission amounts. For this each EU country has to develop a National Allocation Plan (NAP) outlining the total number of emission allowances (EUAs) allocated to the individual installations covered by the scheme.[68] According to the European Commission, the allowances should be in line with the member states' Kyoto commitments. At the end of each year, each site must surrender sufficient allowances to cover their CO_2 emissions for that year. Failure to do so will result in fines - 40 euros per ton of CO_2 in the first period, and 100 euros in 2008-2012. In addition, the deficit must be compensated for in the following year. [69]

The Protocol also establishes a form of emissions trading among Annex I countries that revolves around projects that

---

[68] *Id*, Art 9 provides that for each period referred to in Article 11(1) and (2), each Member State shall develop a national plan stating the total quantity of allowances that it intends to allocate for that period and how it proposes to allocate them. The plan shall be based on objective and transparent criteria, including those listed in Annex III, taking due account of comments from the public. The Commission shall, without prejudice to the Treaty, by 31 December 2003 at the latest develop guidance on the implementation of the criteria listed in Annex III.

[69] *Id*, Art 16(3)&(4) states Member States shall ensure that any operator who does not surrender sufficient allowances by 30 April of each year to cover its emissions during the preceding year shall be held liable for the payment of an excess emissions penalty. The excess emissions penalty shall be EUR 100 for each tonne of carbon dioxide equivalent emitted by that installation for which the operator has not surrendered allowances. Payment of the excess emissions penalty shall not release the operator from the obligation to surrender an amount of allowances equal to those excess emissions when surrendering allowances in relation to the following calendar year.(4) During the three-year period beginning 1 January 2005, Member States shall apply a lower excess emissions penalty of EUR 40 for each tonne of carbon dioxide equivalent emitted by that installation for which the operator has not surrendered allowances. Payment of the excess emissions penalty shall not release the operator from the obligation to surrender an amount of allowances equal to those excess emissions when surrendering allowances in relation to the following calendar year.

reduce or remove emissions, referred to as Joint Implementation ("JI").[70] In a JI transaction, an Annex I Party invests in a project in the country of another Annex I Party, presumably because the cost of achieving such reductions is lower in the host country than in the investing country. The host Annex I Party then transfers a corresponding portion of its assigned amount to the investing Annex I Party in the form of Emission Reduction Units ("ERUs"). The investing Annex I Party can add these ERUs to its assigned amount. ERUs may be earned only for reductions or removals occurring during the 2008-2012 commitment period.[71]

A significant innovation of the Kyoto Protocol is the establishment of the CDM. Through the CDM, Annex I governments and companies or other persons authorized by them can earn certified emission reductions ("CERs") by investing in emission reduction projects in non-Annex I countries.[72] CDM has been the primary mechanism for involvement of developing countries during the Kyoto Protocol's first commitment period. Because of a perceived abundance of low-cost mitigation in developing countries.

The Kyoto Protocol compliance system is more robust than that of any other international environmental agreement and has introduced a number of innovations to international law generally. The Protocol's compliance system includes mechanisms to generate information about performance, mechanisms to facilitate compliance, and mechanisms to deter non-compliance through penalties. To evaluate compliance, Article 8 of the Protocol calls for the establishment of expert review teams. These teams are empowered to audit information submitted by Annex I Parties pursuant to Article

---

[70] Supra note 51 Art 6.

[71] Council Decision 2002/358/EC of 25 April 2002 concerning the approval, on behalf of the European Community, of the Kyoto Protocol to the United Nations Framework Convention on Climate Change and the joint fulfilment of commitments there under OJ L 130, 15.5.2002 P. 0001-0003.

[72] Supra note 51 Art 12.

5 and Article 7. Claims of non-compliance come before the Compliance Committee,[73] which consists of two bodies: the Facilitative Branch and the Enforcement Branch, each consisting of delegates appointed by the Parties. The Facilitative Branch, consistent with its name, has assistance and early warning functions and aims to prevent non-compliance before it occurs. It can direct financial and technical assistance to Parties. The Enforcement Branch, by contrast, has quasi-judicial functions.[74] It assesses compliance by Annex I Parties with respect to reporting requirements and Article 3 commitments.

The KP came into force February 16, 2005 and in the end of 2008 it was already ratified by 176 countries. The Kyoto Protocol took much longer to enter into force that had been expected due to the fact that the Kyoto Protocol only contained very general rules e.g. concerning the mechanisms joint implementation, clean development mechanisms and emissions trading, a compliance mechanism and financial mechanisms. Only at COP-7 in 2001 agreement could be reached on controversial questions, inter alia, on the specific rules for the operation of the mechanisms, the functioning of the compliance mechanism, financial issues. These compromises were reached in Marrakesh, the Kyoto Protocol became "rectifiable" for Annex I Parties, as they knew then the conditions under which they would have to implement their commitments, i.e. the reduction and/or limitation of their emissions.[75]

The adoption of the Protocol strengthened institutional and legal frames of the climate change regime but, at the same time, it created new challenges to the parties, such as the need

---

[73] *Id*

[74] UNFCC,COP-7,2001,The Marrakesh agreement, based on universally agreed texts of more than 200 pages, provides a semblance of finality to the Kyoto Protocol.

[75] *Id* See V Sridhar, The Marrakesh Deal, *Frontline*, Volume 18 - Issue 25, Dec. 08 -21, 2001.

for a detailed composition of flexibility mechanisms and the question of equity between developed and developing countries. Particularly the issue of equality regarding different commitments was the reason that the post-Kyoto negotiations positive attitude of further negotiations started to crack and forecasted more disagreements between the parties. As an initial attempt to address these risks, the Kyoto Protocol's commitment period officially began in January 2008. A number of countries are undertaking substantial actions to limit their emissions. The European Union moved into the second phase of its Emission Trading Scheme (EU ETS), now the world's largest carbon dioxide (CO_2) cap-and-trade program, covering approximately half of all EU CO_2 emissions. Efforts to promote climate-friendly development in low-income countries have delivered projected emissions reductions in excess of 1.75 billion tonnes of CO_2 equivalent through the Clean Development Mechanism (CDM).[76]

After the Kyoto protocol was signed negotiations of COP continued and various compromises were reached. For instance the continued negotiation in COP-4 was concluded with adoption of the Buenos Aires Plan of Action which scheduled timetable for negotiating the "modalities" that would expand principles, guidelines and strategies of the Kyoto Protocol.[77] According to the Plan, advanced rules of the Kyoto flexible mechanisms were to be adopted during the COP-6.[78] Unfortunately, the COP-6 in Hague in 2000 finished with the collapse of negotiations and did not finalize any

---

[76] Joseph Aldy, "Climate Policy Architectures for the Post-Kyoto World", *Journal Article*, environment, volume 50, issue 3, pages 6-17, May, June(2008).

[77] Buenos Aires Plan of Action'COP-4,1998. The Fourth Conference of the Parties to the UNFCCC concluded with the adoption a 'Buenos Aires Action Plan' establishing deadlines for finalizing work on the Kyoto Mechanisms (Joint Implementation, Emissions Trading and the Clean Development Mechanism), compliance issues and policies and measures.

[78] Implementation of the Buenos Aires Plan Of Action, UNFCCC, 2001, Decision 1/CP.6.

package deal or agreement. The COP-6 failure indicated a complexity and serious economic implications awaiting the post-Kyoto negotiations evolution of the regime.[79] Further negotiations continued in November 2001 during the COP-7 when parties eventually finalized rules and procedures of the KP through adoption of the Marrakesh Accords Adoption of Marrakesh Accords marked the end of the legislative era and started the phase of implementation in the climate change regime.[80] Further negotiations reached a landmark in COP-13 meeting in 2008 held in Bali.

BALI ACTION PLAN

In response to the lack of binding commitments for developing countries, the international community finally, in 2005, at COP-11, agreed to start a series of "dialogues" to consider long-term cooperation under the UNFCCC "without prejudice to any future negotiations, commitments process, framework or mandate under the Convention."[81] At Bali in 2008, at COP-13, the Parties to the UNFCCC established the Ad Hoc Working Group on Long-term Cooperative Action (AWGLCA) as a subsidiary body under the UNFCCC. The Bali Action Plan sets out a roadmap that includes four building blocks: mitigation, adaptation, technology transfer, and financial resources. [82]In regards to mitigation, AWGLCA is to consider mitigation commitments or actions, including quantified emission limitations and reduction objectives by all developed country Parties. [83] Developing countries are to have

[79] UNCTAD, Trade and Development Report, 2001, p. 35-37.

[80] Ott, H.E., Climate Policy After the Marrakesh Accords: From Legislation to Implementation. Global Climate. *Yearbook of International Environmental Law*, Vol. 12 Oxford University Press,p 11 (2001).

[81] United Nations Framework Convention on Climate Change, Bali Action Plan, Decision -/CP.13, at 1 (Dec. 15, 2007), available at http://unfccc.int/files/meetings/cop_13/application/pdf/cp_bali_action.pdf (last visited 28-03-, 2009).

[82] *Id*

[83] *Id*

nationally appropriate mitigation actions in the context of sustainable development, supported and enabled by technology, financing, and capacity-building, in a measurable, reportable, and verifiable manner. The Bali Action Plan calls for international cooperation to support urgent implementation of adaptation action using vulnerability assessments, financial-needs assessments, and capacity-building and response strategies, among others.[84] Risk management is promoted as well as insurance to transfer risks.[85]

In regards to technology transfer, AWGLCA is to consider means for removal of obstacles and provisions of incentives to scale-up the development and transfer of technology to developing countries, thus promoting access to affordable, environmentally-sound technology. [86] Enhanced action on the provision of financial resources will be based on consideration of improved access to adequate, predictable and sustainable financial resources for developing countries, including new and additional resources and mobilization of private sector funding and investment. [87]

Bali Action Plan outlined some of the other subjects for the future discussion which include the use of sectoral approaches; approaches to enhance the cost-effectiveness of mitigation actions, including market mechanisms; and the issue of reducing emission from deforestation and forest degradation in developing countries. International cooperation to support urgent implementation of adaptation actions, including through vulnerability assessments, prioritization of actions, financial needs assessments, capacity-building and response strategies, integration of adaptation actions into sectoral and national planning, specific projects

---

[84] *Id*

[85] *Id*

[86] *Id*

[87] *Id*

and programmes, means to incentivize the implementation of adaptation actions, and other ways to enable climate-resilient development and reduce vulnerability of all Parties, taking into account the urgent and immediate needs of developing countries that are particularly vulnerable to the adverse effects of climate change, especially the least developed countries and small island developing States, and further taking into account the needs of countries in Africa affected by drought, desertification and floods.[88]

In Bali it was agreed that the post-2012 arrangement concerning climate change is to be agreed at COP-15 and COP/MOP5, which is to be held in Copenhagen. It was for COP-14 and COP/MOP4 in Poznan/Poland to evaluate the progress made in the various fora and to propose the work-programme until December 2009. The meetings at Poznan were an important half-way mark in the negotiating process leading up to Copenhagen in 2009. The delegates welcomed the determination of the Ad Hoc Working Group on Long-term Cooperative Action under the Convention to shift into full negotiating mode in 2009. Poznan strategic programme on technology transfer was announced as a step towards scaling up the level of investment in technology transfer in order to help developing countries address their needs for environmentally sound technologies.[89] Developed country Parties were called upon and other Parties were invited to make financial contributions to the Global Environment Facility(GEF). Specific provisions were envisaged regarding GEF. It urged the Council of the Global Environment Facility to help to ensure that adequate and predictable funding is made available for the implementation of GG reduction commitments under the Convention.[90]

---

[88] Advancing the Bali Action Plan, Decision 1/CP.14.

[89] Development and transfer of technologies, UNFCC 2009 Decision 2/CP.14.

[90] Financial mechanism of the Convention: fourth review of the financial mechanism, UNFCC 2009 Decision 3/CP.14.

It empowered the Global Environment Facility to fully address the various issues raised over the implementation of the Resource Allocation Framework; and to continue action on mitigation and adaptation measures by appropriate, transfer of, or access to, environmentally sound technologies and know-how. Further emphasis has been granted to improve access for all developing countries, in particular least developed countries, Small Island developing States and countries in Africa, to Global Environment Facility resources.[91]

It envisages on operating the Least Developed Countries Fund which will strive to work with its agencies to improve communication with least developed country and improve access funding and other support for the preparation and implementation of projects identified in national adaptation programmes of action. Further the fund will be having the responsibility of augmenting collaboration with its agencies and the Least Developed Countries Expert Group, the remaining least developed country Parties that have not submitted their national adaptation programmes of action, in completing and submitting their national adaptation programmes of action as soon as possible.[92] Stress was laid on capacity building by developing and least developed nations.

THE GLOBAL ENVIRONMENT FACILITY: FINANCIAL MECHANISM

Any discussion on the climate change legal regime will not be complete without an analysis of Global Environment facility. The Global Environment Facility (GEF) is a global partnership among 178 countries, international institutions, non-governmental organizations (NGOs), and the private

---

[91] Additional guidance to the Global Environment Facility, UNFCC 2009 Decision 4/CP.14.

[92] Further guidance for the operation of the Least Developed Countries Fund, Decision 5/CP.14 Report of the Conference of the Parties on its fourteenth session, held in Poznan from 1 to 12 December 2008.

sector to address global environmental issues while supporting national sustainable development initiatives. It provides grants for projects related to six focal areas: biodiversity, climate change, international waters, land degradation, the ozone layer, and persistent organic pollutants.[93]

It was established in October 1991 as a $1 billion pilot program in the World Bank to assist in the protection of the global environment and to promote environmental sustainable development. The GEF provide new and additional grants and concessional funding to cover the "incremental" or additional costs associated with transforming a project with national benefits into one with global environmental benefits.[94] In 1994 at the Rio Earth Summit, the GEF was restructured and moved out of the World Bank system to become a permanent, separate institution. Today the GEF is the largest funder of projects to improve the global environment. Since 1991, GEF has achieved a strong track record with developing countries and countries with economies in transition, providing $8.26 billion in grants and leveraging $33.7 billion in co-financing for over 2,200 projects in over 165 countries.[95] The GEF is also the designated financial mechanism for a number of multilateral environmental agreements (MEAs) or conventions including UNFCC. As the financial mechanism of the UNFCCC, GEF allocates and disburses about $250 million dollars per year in projects in energy efficiency, renewable energies, and sustainable transportation. Moreover, it manages two special funds under the UNFCCC – the Least Developed Countries Fund and the Special Climate Change Fund.

---

[93] Global Environment Facility available at *http://www.gefweb.org/interior_right.aspx?id=50(last accessed on 4-03-2009)*

[94] *Id*

[95] *Id*

The GEF supports projects in:

Climate Change Mitigation: Reducing or avoiding greenhouse gas emissions in the areas of renewable energy, energy efficiency, and sustainable transport.

Climate Change Adaptation: Aiming at increasing resilience to the adverse impacts of climate change of vulnerable countries, sectors, and communities.[96]

CONSTRUCTING A POST 2012 CLIMATE CHANGE REGIME: CHALLENGES FOR THE INTERNATIONAL LAW

Kyoto protocol aimed at bringing down the GHG emissions by 1990 levels and imposed reduction commitments Annex 1 countries between 2007-2012. The question always persisted what after this commitment period. The history and implementation of Kyoto protocol throws open certain disturbing questions. In order to bring an equitable climate change regime policy makers have to think a fresh and rectifying various drawbacks of the existing mechanisms.

A persistent criticism of the Kyoto Protocol is its alleged failure to include so-called developing countries under its emission control system. Most glaringly, the agreement's emission control system does not cover fast-growing countries like China, India, and Brazil. From an environmental standpoint, it is simply not possible to protect the climate system over the long term if developing countries do not participate in emission reductions.[97] Nearly 80 per cent of the world's population lives in developing countries, which already account for almost 50 per cent of the world's GHG

---

[96] *Id*

[97] Nakicenovic, Nebojsa and Swart, Robert, eds. Special report on emissions scenarios: *A* special report of working group III of the Intergovernmental Panel on Climate Change. Cambridge: Cambridge University Press, 2000.

emissions.[98] The decision to not impose reduction obligations on countries with the capacity and potential to emit large quantities of GHG in the future means that, regardless of whether Annex I countries comply with their obligations, overall GHG emissions worldwide may continue to rise on account of the emissions from those potentially future large emitters who are presently not placed under reduction obligations. [99] If this happens, the achievement of the ultimate objective of the UNFCCC and Kyoto would be defeated. To make approval of developing countries to commit to GHG reduction will remain as the major challenge for the ale makers. On the one hand, for purposes of environmental effectiveness and economic efficiency, key developing countries should participate. On the other hand, for purposes of distributional equity (and international political pragmatism), they cannot be expected to incur the consequent costs.[100]

First of all, the United States and Australia, the two serious GHG emitters (per capita), refused to ratify the Protocol and, thus, are not bound to any obligations regarding mitigation efforts as well as cannot participate in the emissions trading within the flexible mechanisms. Secondly, the South continues to resist any substantial commitments under the Convention, and probably will not change such position since it prioritizes the right to continue its economic development as well as expects that industrialized nations will meet their targets. Thirdly the Kyoto Protocol, with its five year time horizon 2008 to 2012, represents a relatively short-term approach for what is fundamentally a long-term problem. Greenhouse

---

[98] Kevin Baumert & Jonathan Pershing, Climate Data: Insights and Observations 37 (2004).

[99] Richard B. Alley et al., Summary for Policymakers, in Intergovernmental Panel on Climate Change, 2007, The Physical Science Basis 2, 18 (2007).

[100] Supra note 65 See Hann.

gases have residence times in the atmosphere of decades to centuries.[101] Furthermore, to encourage the magnitude of technological change that will be required to meaningfully address the threat of climate change it will be necessary to send long-term signals to the private market that stimulate sustained investment and technology innovation.[102] Finally, the Kyoto Protocol may not provide sufficient incentives for countries to comply. [103] Some countries' emissions have grown so fast since 1990 that it is difficult to imagine that they can undertake the emission mitigation or muster the political will and resources to purchase enough emission allowances or CDM credits from other countries to comply with their targets under the Protocol.

There are various concerns regarding the CDM. This is not a cap-and-trade mechanism, but rather an emissions reduction- credit system. That is, when an individual project results in emissions below what they would have been in the absence of the project, a credit—which may be sold to a source within a cap-and-trade system—is generated. This approach creates a challenge: comparing actual emissions with what they would have been otherwise. The baseline—what would have happened had the project not been implemented— is unobserved and fundamentally unobservable. In fact, there is a natural tendency, because of economic incentives, to claim credits precisely for those projects that are most profitable, and hence would have been most likely to have been executed without the promise of credits. This so-called "additionality problem" is a serious issue. [104]

---

[101] Newell, The political economy of global environmental governance' *Review of International Studies* July (2008).

[102] *Id*

[103] Scott Barrett, "Rethinking Global Climate Change Governance". *Economics Discussion Papers*, (2008).

[104] Robert N. Stavins, Getting serious about Global Climate Change: post-Kyoto international Climate Policy Architecture, available at

Contd....

CONCLUSION

Human-induced climate change will affect the future of the planet and human life in numerous ways. This justifies the need for climate change reduction and mitigation as the core principle of any legal regime seeking to address this problem.[105] The threat of global climate change is exceptional. As acknowledged by the Intergovernmental Panel on Climate Change (IPCC), it has "the potential to lead to future large-scale and possibly irreversible changes in Earth systems resulting in impacts at continental and global scales."[106]

The goal of the international community must be to move much more quickly to act on climate change, especially now that we know that we do not have decades to figure out what to do. [107] If we analyse the Bali Roadmap it does not specify what form of an agreement is supposed to be adopted in Copenhagen in 2009. Will there be a new framework convention with new goals on global emission reductions--long-term and short-term?As the international community deals with climate change over the next several decades, it will face many issues of justice, equity, and ethics. For example, already there are people whose lives and cultures are being disrupted or destroyed by climate change.[108] Some of the steps

---

Contd. from previous page...

http://www.pc.gov.au/__data/assets/pdf_file/0006/86748/12-chapter 9. pdf (Last accessed on 12-03-2009).

105 As noted by the Intergovernmental Panel on Climate Change, most of recent warming is attributed to human activities. See Intergovernmental Panel on Climate Change, Climate Change 2007: The Physical Science Basis, Summary for Policymakers 8 (2007).

106 *Id*

107 Press Release, United Nations Foundation, Science Panel Outlines Roadmap for Reducing Risks from Climate Change, (Feb. 27, 2007), available at http://www.unfoundation.org/media_center/press/2007/pr_22707.asp. (Last accessed on 2-04-2009).

108 For a description of the impact of climate change on the Inuit peoples of the far north, see Petition to the Inter American Commission on Human Rights Seeking Relief from Violations Resulting from Global Warming

Contd....

which will substantially address these concerns are dealt below.

The best result that can be hoped for to result from the COP-15 in Copenhagen in 2009 is to amend the UNFCCC and the Kyoto Protocol to get near universal participation in a more forceful way to address the enormous challenges of climate change, so every nation feels its concerns are addressed. The Annex III Parties would then have commitments under the Kyoto Protocol under a new "Annex C". This is because participation of developing countries in mitigation efforts is the future key element and the most controversial issue of the climate change regime especially that emissions of industrializing and populated countries, such as China and India, are estimated to grow and soon will require some sort of a control and form of reduction as well. On the other hand, much more vulnerable nations of the South will also have to face their exposure and weakness towards various impacts of the climatic changes and, thus, perhaps they will have to force adaptation strategies through on the negotiations' agenda under the UNFCCC. To complement Annex C, a new Annex C Mitigation Fund needs to be incorporated into the Kyoto Protocol as "Article 11-Bis," placed immediately after Article 11, the financial provision. This fund would have to be substantial to help the Annex III Parties receive technology transfers and other assistance much faster than provided by the current financial mechanism under the Climate Convention and the Kyoto Protocol that is operated by the Global Environment facility. Following the trend in the U.N. to partner with the private sector, the Annex C Mitigation Fund would not necessarily be run by the Global Environment Facility, but could be managed by an international private bank accountable to and guided by the

---

Contd. from previous page...

Caused by Acts and Omissions of the United States (Dec. 7, 2005), available at http://www.inuitcircumpolar.com/files/uploads/icc-files/FINAL Petition ICC.pdf hereinafter Inuit Petition.

COP/MOP. The goal would be to use the more efficient private banking system to track, manage, and disperse the funds earmarked for the financing of the incremental costs of "clean" technology transfer and other assistance to Annex C countries. Regarding the Kyoto Mechanisms, Annex C Parties would be able to take advantage of all the flexible mechanisms. Currently, developing countries can only participate in the Clean Development Mechanism (CDM). Participating in all three mechanisms is important because being eligible to participate in emissions trading can help Annex III Parties economically, while at the same time fulfilling their reduction commitments.[109]

Further poor countries' position at negotiations for a post Kyoto protocol should be that the allocation of emission entitlements should be based on the sustainable development principle of intra-generational equity, i.e., equity among the present generations. Intra-generational equity requires that the world's resources--in this case GHGs emission rights--be allocated equitably among all nations since the emission entitlements, in effect, are a proxy for the right to develop and meet the needs of one's nation and the well-being of its people.[110] Poor countries' position at the post Kyoto regime negotiations must be that the basis of equitable allocation should be the Human Development Index developed by the U.N. Development Program (UNDP) [111] combined with historic GHG emissions, energy use increases needed to improve low HDI and associated Human Poverty indices, and the efficiency with which energy is used. The Human

---

[109] Kevin A. Baumert, "Participation of Developing Countries in the International Climate Change Regime: Lessons for the Future, *38 GEO. WASH. INT'L L. REV. 365*, 390-91 (2006).

[110] Albert Mumma, "Designing a Global Post Kyoto Climate Change Protocol that advances Human Development, *20 Geo. Int'l Envtl. L. Rev*,619,(2008).

[111] U. N. DEV. PROGRAMME, HUMAN DEVELOPMENT REPORT 2007/ 2008 111 (2007) .

Development Index is more nuanced than the principle of per capita emissions, which currently dominates the discussions regarding the formula for assigning responsibility for mitigating climate change. Beyond the question of dealing appropriately with climate change victims, the international community must consider the nature of its obligations to future generations. Is there an obligation for living humans to preserve or improve the earth's condition for future humans?[112]

Further climate change-induced damage must be addressed simultaneously at the national and international levels. We should think seriously in establishing a legal regime based on liability. Already at national level and to an extent at the international level the principle of state responsibility exists in certain environment matters like transboundary pollution. The points that can be advocated in favour of establishment of an international liability principle in climate change are as follows. Indeed, any damage that occurs in a given country as a result of climate change can often be ascribed to past and/or present activities taking place in other parts of the world. Further, any liability regime limited to a national level framework would in most cases deny claimants access to compensation. Indeed, only a few countries and private actors have been and are responsible for most greenhouse gas production or emission. Limiting liability to the national level would deny compensation to all people in countries where no liable entity is found. States are responsible for the consequences of breaches of international law.[113] Yet, even though this principle is largely accepted, there is no binding international legal regime concerning state

---

[112] Jonathan C. Carlson, "Reflections on a Problem of Climate Justice:Climate Change and the Rights of States in a Minimalist International legal Order", *18 Transnat'l L. & Contemp. Probs*,45,(2009).

[113] Draft Articles on Responsibility of States for Internationally Wrongful Acts Art. 1, International Law Commission, G.A. Res. 56/83, Annex, U.N. Doc. A/56/10(SUPP) (Dec. 12, 2001).

responsibility. States have usually preferred to use other mechanisms to solve their disputes The lack of state enthusiasm for the development of a comprehensive state responsibility regime at the international level to address environmental damage, combined with the fact that only states can bring claims, helps explain in part the importance that civil liability regimes have acquired in the field of environmental law. The liability and redress regime that needs to be adopted should be international and comprehensively address all issues related to climate change. In fact, it is difficult to conceive of addressing liability effectively any other way. This is because most problems that will occur involve an extraterritorial element. In certain situations affected individuals may be able to bring actions against their own government or companies for the damages suffered.[114] The first hurdle in the development of a liability regime for global warming is the need for certain countries and economic actors to acknowledge their responsibility in causing the phenomenon, something developed countries have refused to do until now.[115]

To maximize the chances of reducing global GHG levels, climate change parties should develop regional agreements regarding the building of major-emitting industrial plants. Under this scheme, regional parties could provide input and be put on notice about projected emissions due to industry growth. Regional agreements with climate change parties can address more limited goals and sub-issues of the threat. The parties at the climate change negotiating table could develop greater incentives to cooperate by responding to limited issues with regional agreements. They could better use trade and

114 On the potential link between hurricane Katrina and global warming, see, e.g., Pew Center on Global Climate Change, Katrina and Global Warming, available at http://www.pewclimate.org/special reports/katrina.cfm.(Last accessed on 6-01-2009).

115 Philippe Cullet, "Liability and Redress for Modern Biotechnology", *15 Y.B. Int'l Envtl. L.* 165 (2004).

economic bargaining approaches, especially with the transportation sector, and better promote the effectiveness of treaties, while simultaneously addressing the sub-issues of climate change.[116] For example, in 2005, the "coal pact" countries, including the United States, China, Australia, Japan, India, and South Korea, through the Asia Pacific Partnership on Clean Development and Climate, agreed to promote technologies to reduce carbon dioxide emissions in coal and implement cleaner burning.[117]

Achieving long-term climate change policy goals will require a ramp-up in the innovation and deployment of energy-efficient and low-carbon technologies. Thus, a suite of policies is needed to provide the proper incentives for technological change. Three categories of policies are potentially important: international carbon markets, technology transfer to developing countries, and coordinated innovation and commercialization programs. Beyond putting a price on carbon, the next international climate agreement can provide several mechanisms to facilitate the development and deployment of climate-friendly technologies. Examples include providing a venue for countries to pledge resources for technology transfer and R&D and coordinating agreement on principles for allocating resources. Likewise, barriers to the transfer of climate-friendly technologies could be reduced through a World Trade Organization (WTO) agreement that lowered tariff and non-tariff barriers to trade in environmental goods and services.

---

116 Lisa Schenck, "Climate Change Crisis" - Struggling for Worldwide Collective Action, *19 Colo. J. Int'l Envtl. L. & Pol'y* 319,(2008).

117 White House Climate Change Initial Report 1 (2001), available at www.whitehouse.gov/news/releases/2001/06/climatechange.pdf.(Last accessed on 23-03-2009).

2

National Action Plan on Climate Change

India's Legal Response to Global Warming

INTRODUCTION

India has always been in the forefront of climate change discourse internationally and at the domestic level. Number of technological and policy measures were initiated aimed at reducing the green house gas emissions which is responsible for climate change. But we had been shying away from drafting any direct policy framework on climate change. India has been under tremendous pressure from several quarters over its climate change policy. The recent National Action Plan on Climate Change (NAPCC) released by the government of India tried to overcome these criticisms. The NAPCC has laid down a national strategy on climate change with regard to adaptation and mitigation of greenhouse gas emissions, both domestically and in the context of global arrangements under the United Nations Framework Convention on Climate Change (UNFCCC). Sustainable development is the key theme running through the breadth of the plan. The plan in a way marks a new direction in India's climate policy. It brings about a more coherent and systematic approach to climate changes and thereby aims to achieve sustainable development. The plan will definitely give a boost to India's suitability at the global level to augment our bargaining control with the developed nations, especially for providing clean technologies at cheaper rates. At the same time certain serious shortcomings in the plan raises various questions marks over the efficacy of the plan. The chapter critically analyse India's various policy response to Climate change. It gives special emphasis to NAPCC and sees how far this plan will be instrumental in bringing sustainable development without compromising our development needs.

INDIA AND CLIMATE CHANGE

India house a population of 1 billion plus. Its economy is heavily dependent on its diversified natural resources and various climate dependent sectors including agriculture, forestry, and energy sector. With such a heavy economical dependence on natural resources India faces a severe peril because of the anticipated climate change. Though it can not be predicted with certainty the possible impact of climate change in India, various studies have pointed out that certain critical sectors can be negatively affected by the climate change. The following impacts have been identified by the Government of India's initial National Communications (NATCOM I) in 2004.[1] It reports that Climate change is likely to impact all the natural ecosystems as well as socio-economic systems.[2] This latest high resolution climate change scenarios and projections for India, based on Regional Climate Modelling (RCM) system, known as PRECIS developed by Hadley Center and applied for India using IPCC scenarios indicates the following.[3]

---

[1] India's Initial National Communications to the United Nations Framework Convention on Climate Change, Ministry of Environment and Forests, New Delhi, (2004).

[2] *Id*

[3] Rupa Kumar, K., "High-resolution climate change scenarios for India for the 21st century". *Curr. Sci.*, 90, p 334-345, (2005).

TEMPERATURE

At the global level an increase of 0.4° C has been registered in surface air temperatures over the past century. With regard to India marked increase in seasonal surface air temperature is projected into the 21st century, becoming conspicuous after the 2040s.[4] Hadley Center Regional Model (Had RM2) and the IS92 projections indicate an increase in minimum temperature by 4°C all over the country.[5]

PRECIPITATION PATTERNS

The pattern of average rainfall at the national level during the past century does not show a clear trend. But at the regional level an increasing monsoon seasonal rainfall has been found along the west coast, northern Andhra Pradesh and north-western India (+10% to +12% of the normal over the 100 years) while a trend of decreasing monsoon seasonal rainfall has been observed over eastern Madhya Pradesh, north-eastern India, and some parts of Gujarat and Kerala (-6% to -8% of the normal the 100 years).[6] Concentration of droughts is projected in Gujarat and Rajasthan, which are already drought-prone, and in Orissa, which is currently flood-prone.[7]

SEA LEVEL RISE

It is expected that rise of sea level is going to cause irreversible damage to coastal areas displacing Lakhs of people, affecting coastal eco systems and agriculture. A mean sea level rise of 15-38 cm is projected along India's coast by

---

4 *Supra* note 1.

5 *Id*

6 C V Naidu, B R Srinivasa Rao and D V Bhaskar Rao, "Climatic Trends and Periodicities of Annual Rainfall over India, Meteorological Applications", Cambridge University Press, 6:395-404 (1999).

7 Climate Change and India: Impacts, Policy Responses and a Framework for EU-India Cooperation, DG Internal Policies of the Union Policy Department Economic and Scientific Policy.

the mid- 21st century and of 46–59 cm by 2100. [8] India's NATCOM I assessed the vulnerability of coastal districts based on physical exposure to Sea Level Rise, social exposure based on population affected, and economic impacts. Coral reefs may decline owing to rise in sea surface temperatures. It is estimated that if a one-meter sea level rise were to take place today, it would displace 7 million persons in India.[9]

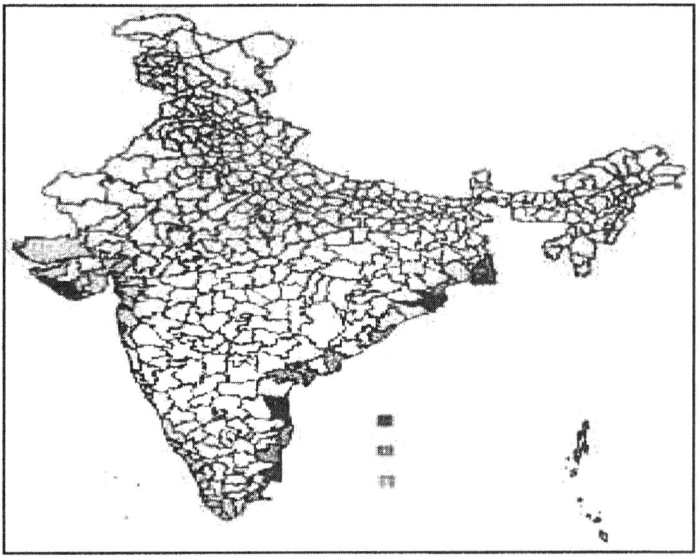

Figure 2·1: Sea level Rise projections

Source: NATCOM

IMPACT ON WATER RESOURCES

The water cycle, a fundamental component of climate, is likely to be altered in important ways due to climate change. In India most significant changes are likely to occur in the

---

[8] *Supra* note 1

[9] Jyoti K. Parikh and Kirit Parikh, Climate Change: India's Perceptions, Positions, Policies and Possibilities, Climate Change and Development, OECD Discussion paper, (2002).

glacier fed river systems. The Intergovernmental Panel on Climate Change (IPCC)[10] predicts that glacial melts in the Himalayan region alone will increase flooding, trigger avalanches and landslides, and cause extinction of species and ecosystems. [11]This will substantially change the long-term water flows of snow- and glacier-fed river systems; of the Brahmaputra, the Ganga, and the Indus. As such, Himalayan glaciers, including the Gangotri, which is a source of the perennial and holy river Ganga, have receded by thirty meters, endangering water supply in the dry season. [12] River basins of Sabarmati and Luni, which occupy about one quarter of the area of Gujarat and 60 per cent of the area of Rajasthan, are likely to experience acute water scarce conditions.[13] Same goes with river basins of Mahi, Pennar, Sabarmati and Tapi. Further rising sea levels may lead to increased saline invasion into coastal and island aquifers, while increased incidence and severity of floods may affect groundwater quality in alluvial aquifers.

---

[10] The Intergovernmental Panel on Climate Change, comprising scientific experts from a select number of countries, was established in 1988 by the Executive Council of the World Meteorological Organization (WMO), with support from UNEP, to identify "uncertainties and gaps in "present knowledge" about climate change and its potential impacts; gather information "to evaluate policy implications" and "response strategies"; "review... national/international policies" regarding GHGs; and carry out "scientific and environmental assessments regarding GHG emissions and transfer them for relevant governmental and intergovernmental agencies.

[11] Climate Change 2007 Impacts, Adaptability and Vulnerability, IPCC, Working Group II Contribution to the Intergovernmental Panel on Climate Change Fourth Assessment Report, Summary for Policy Makers (2007), available at http://www.ipcc.ch/pdf/assessment-report/ar4/wg2/ar4-wg2-spm.pdf.

[12] Emily Wax, A Sacred River Endangered by Global Warming, Wash. Post, June 17, 2007, at A14, available at http://www.washingtonpost.com/wp-dyn/content/article/2007/06/16/AR2007061600461.html.

[13] Sustainable Development and Climate Change a business outlook, BACKGROUND PAPER, TERI Business Council for Sustainable Development India.

IMPACT ON AGRICULTURE AND LIVELIHOODS

The most serious and immediate impact of climate change is predicted on agriculture. Various studies have indicated a probability of 10 to 40 per cent loss in crop production in the country due to the anticipated rise in temperature by 2080-2110.[14] Studies conducted by the New Delhi-based Indian Agricultural Research Institute (IARI) have pointed to a possible loss of 4 to 5 million tonnes in the overall wheat production with every 1 degree centigrade increase in temperature throughout the growing period of the crop.[15] It is also predicted that Wheat production for the country as a whole may decline after 2020 and rice production may be adversely impacted in the eastern states.

In a more detailed study of India, Kumar and Parikh examined the impact of climate change on agricultural crop yields, GDP and welfare. They estimated that: (a) without considering the carbon dioxide fertilization effects yield losses for rice and wheat vary between 32 and 40%, and 41 and 52%, respectively.[16] Their study also showed that even with carbon fertilization effects, losses would be in the same direction but somewhat smaller. Using an alternative methodology Kumar and Parikh showed that even with farm-level adaptations, the impacts of climate change on Indian agriculture would remain significant.[17] They estimated that with a temperature change of +2°C and an accompanying precipitation change of +7 %, farm level total net-revenue would fall by 9%, whereas with a temperature increase of +3.5°C and precipitation change of

---

14 Surinder Sud, Agriculture may decline due to climate change, Business Standard, New Delhi January 21, (2008).

15 *Id*

16 Kumar, K.S.Kavi, Parikh, J., 'Indian Agriculture and Climate Sensitivity', *Global Environmental Change*, 11, p. 147-154, (2001).

17 *Id*

+15%, the fall in farm level total net-revenue would be nearly 25 %.[18]

Other impacts on agricultural and related sectors include lower yields from dairy cattle and decline in fish breeding, migration, and harvests. Small changes in temperature and rainfall have significant effects on the quality of fruits, vegetables, tea, coffee, aromatic and medicinal plants, and basmati rice. For a developing country, these are very large changes which can cause much human misery. From India's point of view, a 2°C increase would be clearly intolerable. Other developing countries may be even more vulnerable possibly Bangladesh or Small Island States.[19]On the positive side, the higher concentration of carbon dioxide in the atmosphere is expected to be beneficial for several crops. For instance damage to the crops due to ground frost, which is quite common during the rabi season in the country's key north-western agricultural belt, may also reduce due to rise in temperature in the wake of climate change.[20] The impact of climate change on agriculture. (Fig. 2·2)

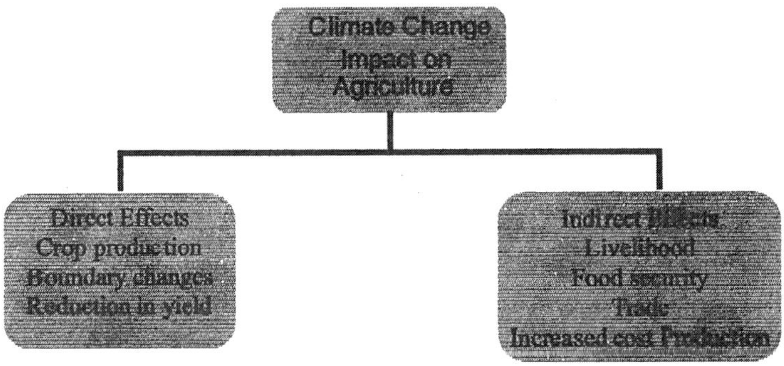

Figure 2·2: Impact of Climate change on Agriculture

Source: Author

[18] Kumar, K.S.Kavi, Parikh, J., 'Socio-economic Impacts of Climate Change on Indian Agriculture', *International Review for Environmental Strategies,* 2(2), (2001).

[19] *Supra* note 7.

[20] Climate Change Impact of Agriculture in India, IARI India.

HEALTH IMPACTS OF CLIMATE CHANGE

Changes in climate may alter the concentration and circulation of many vector species and may augment the spread of diseases to new areas and locations. An increase of 3.8 °C in temperature and a 7% increase in relative humidity by 2050 over present levels is projected to lead to the transmission windows being open for all 12 months in 9 states in India.[21] The transmission windows in the states of Jammu and Kashmir and Rajasthan may increase by 3–5 months. However, in Orissa and some southern states, a further increase in temperature is likely to cut the transmission window by 2–3 months. It is projected that malaria will move to higher latitudes and altitudes in India, with 10 per cent more area offering climatic opportunities for the malaria vector to breed throughout the year during the 2080s.[22] These predictions also indicate the corresponding burden on economy as more financial and infrastructural allocations will be required to tackle the scenario.

VULNERABILITY TO EXTREME EVENTS

Heavily populated regions such as coastal areas are exposed to climatic extremes and large falls in sown areas in arid and semi-arid zones, of which nearly two-thirds are drought-prone. Large areas in Rajasthan, Andhra Pradesh, Gujarat, and Maharashtra and comparatively small areas in Karnataka, Orissa, Madhya Pradesh, Tamil Nadu, Bihar, West Bengal, and Uttar Pradesh are frequented by drought. About 40 million hectares of land is flood-prone, including most of the river basins in the north and the north-eastern belt

---

[21] Ulka Kelkar and Suruchi Bhadwal, South Asian Regional Study on Climate Change Impacts and Adaptation: Implications for Human Development, Paper prepared by TERI for Human Development Report 2007.

[22] *Supra* note 1.

affecting about 30 million people on an average each year. In addition, a projected increase in the intensity of tropical cyclones by 15% poses a threat to the heavily populated coastal zones in the country.[23] In order to make matters worse increased incident of extreme events due to climate change will affect the poor most. For instance in the cyclone in Andhra Pradesh in India in 1996, more than 1,000 people died and there was huge property loss. But major brunt of this was faced by the poor. Cyclones of similar intensity in advanced countries like the U.S. may not lead to any deaths and much hardship, due to stable and durable housing and other infrastructure and extended safety net available to the people in distress.[24]

IMPACTS ON FORESTS AND BIODIVERSITY

Forest plays a major role in balancing GHG emissions. Large scale shifting of forest biomes is indicated as a consequence of climate change. Though in some cases this may result in some positive developments but in most of the cases it will cause irreversible harm in the form of affecting livelihoods of forest dependent communities who use forest, resources for fuel wood, building materials as well as impacting trade of forest products, water supply, and energy production. It may lead to the extinction of various key species seriously affecting the biodiversity of forest ecosystem.[25]

Various studies have been undertaken to analyse the possible consequences of Climate change on forest. Climate impact assessments using BIOME-3 model and climate projections for the year 2085 show 77% and 68%of the forested

---

23 *Id*

24 *Supra* note 9

25 *Id*

grids in India are likely to experience shift in forest types. Indications show a shift towards wetter forest types in the northeastern region and drier forest types in the northwestern region in the absence of human influence. [26]These points are only an indication of the climate change. Vulnerability and impact in India assume high degree of uncertainty in the assessment due to limited understanding of many critical processes in the climate system, existence of multiple climatic and non-climatic stresses, regional-scale variations and nonlinearity.[27]

INDIA: GHG EMISSION PROFILE

India is currently one of fastest growing emitters of green house gasses (GHG) with a 65% increase in emissions between 1990-2005 and projected increase of another 70% by 2020. India is presently the fifth largest GHG emitter in absolute terms and contributes 5% of global emissions.[28]

But India maintains that Climate change is a creation of the blind development strategy followed by west. In support of its claim various statistics are advanced. Indias first NATCOM provides that in 1994, 1,228,540 Gg of CO_2-eq of anthropogenic greenhouse gases (GHGs) were emitted from India resulting in a per capita emission of about 1.3 tons. CO_2 emissions were the largest at 793,490 Gg, i.e. 65 per cent of the total national CO_2-eq emissions. The shares of CH_4 and N_2O were 31 per cent (18,082 Gg) and 4 per cent (178 Gg), respectively (Fig. 2·3).

---

[26] Ravindranath, N. H., Joshi, N. V., Sukumar, R. and Saxena, A., Impact of climate change on forests in India. *Curr. Sci.*, 90,354–361,(2005).

[27] Shukla, P. R., Sharma, S. K., Ravindranath, N. H., Bhattacharya, S. and Garg, A. (eds), *Climate Change and India: Vulnerability Assessment and Adaptation*, Universities Press, Hyderabad, (2003).

[28] Data from IEA, 2007. "CO_2 Emissions from Fossil Fuel Combustion 1971-2005"; IEA, 2007. "World Energy Outlook 2007.

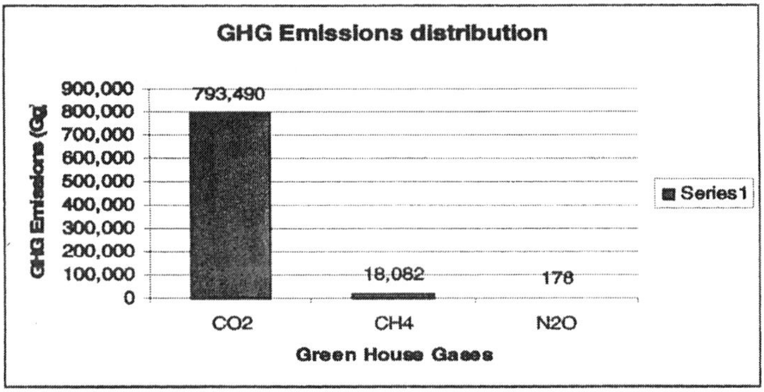

Figure 2·3: GHG Emission India 1994

Source: Author, Data NATCOM

Of the total CO_2-eq emissions in 1994, the largest share of 61 per cent was contributed by the all energy sector, followed by the agriculture sector at 28 per cent. Industrial process at 8 per cent, waste at 2 per cent and land use, land use change and forestry at 1 per cent[29] (Fig. 2·4).

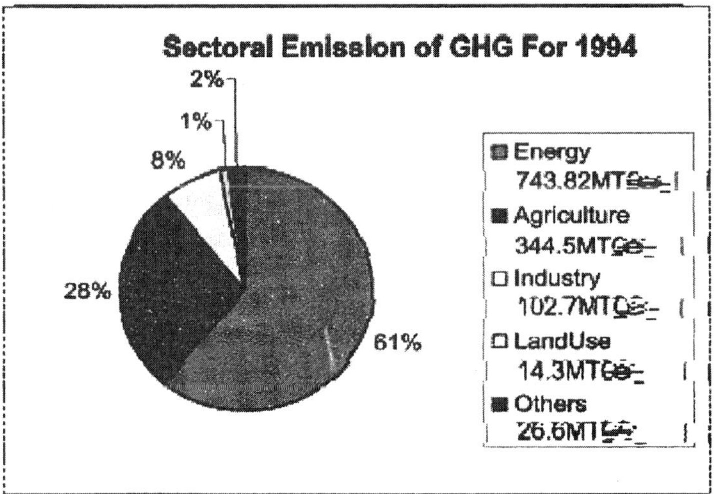

Figure 2·4: Sectoral Emission of GHG 2004

Source: Author, Data NATCOM

[29] *Supra* note 1

But in terms of per capita emission the contribution of India to the cumulative global CO_2 emissions from 1850 and 2000 is only 2%. While the United States of America contributed 30% and the European Union 27% (Fig. 2·5)

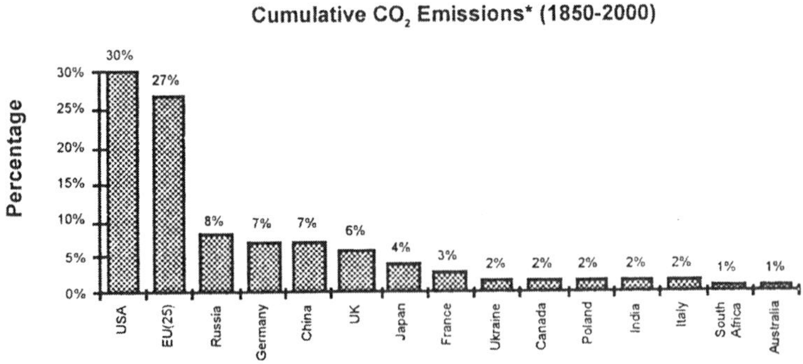

*Energy-related CO_2 only

Source : Climate Analysis indicators Tool (CAIT) version 5.0 (Washington, DC: World Resources Institute, 2008)

Figure 2·5 : Cumulative CO_2 Emissions (1850-2000)

Thus India argues that historically and at present our share in the carbon stock in the atmosphere is relatively very small when compared to the population. India's carbon emissions per person are twentieth of those of the US and a tenth of most Western Europe and Japan. The Indian position on this issue has been quite rigid and consistent. The Government of India has long argued that while climate change is a pressing global problem that needs to be solved, for developing countries like India, the prime goal is that of development and securing a good standard of living for the large number of poor and vulnerable peoples. India has played a insignificant role in bringing about the present problem, which they say, is caused by emissions of the developed countries during their process of industrialization. Thus, India feels that the burden of solving the problem should also rest with these developed countries of the West.

But there are certain factors which are of special concern to India and we can not keep on blaming the west. Of particular concern are the demand trends in the urban areas--electricity for commercial and residential buildings for air conditioners and light, petroleum for increasing private transportation, and coal for cement and steel industries. Infrastructure projects such as road building and electricity generation and distribution will also cause adverse land use shifts.[30] The close link between India's economic growth, infrastructure deficits, energy requirements, and climate consequences translates into complex law and policy-making problem. Unless government policies and legislation balance the competing and overlapping interests, it may not be in a position to sustain its economic growth or prevent climate-related catastrophes.

CLIMATE CHANGE LEGAL REGIME INDIA

India is one of the few developing countries to have incorporated into its Constitution the specific provisions for environmental protection. Article 48A of the Constitution of India provides that 'the State shall endeavour to protect and improve the environment and to safeguard the forests and wild life of the country'.[31] Similarly, Article 51A (g) makes it obligatory for every citizen of India, 'to protect and improve the natural environment including forests, lakes, rivers and wild life, and to have compassion for living creatures.[32] Besides these provisions various entries of constitution empowers the governments to come out with concrete legislation to protect and preserve environment.[33]

With regards to policies, in pursuance of its constitutional mandate various policy and legislative steps were taken by India and has in place many environmental laws including Air Act 1972, Water Act 1974 and Environment Protection Act

---

[30] Jayant Sathaye, P.R. Shukla and N.H. Ravindranath," Climate Change, Sustainable Development and India: Global and National Concerns, 90 *CURRENT SCIENCE*, No. 3, February 10, (2006).

[31] Art 48A,Constitution of India.

[32] *Id*, Art 51 A(g).

[33] V11 Schedule, Constitution of India.

1986 etc., which attempt to reduce pollution and environmental damage, even though not all were framed specifically to tackle climate change. In regard to climate change India's contributions to greenhouse gas emissions are very small, compared to other developed countries. Nevertheless, the administration has undertaken several voluntary measures that will slow down GHG emission rates. These measures include promotion of renewable energy [34] and investments in clean development technologies. India's major emphasis on climate change policies has been in relation to energy conservation.

ENERGY CONSERVATION AND PROMOTION OF RENEWABLE SOURCES

India has one of the most active renewable energy programmes besides having a dedicated Ministry for non-conventional energy sources. India has for quite some time pursued energy preservation policies in her own interest. India's obligation to minimize energy consumption has been majorly attempted through various electricity Acts. The major emphasis of these electricity Act has been to conserve and generate more electricity through scientific practice. Another prominent policy initiative has been to price energy at a high level in PPP terms through taxes, duties, and cess. Standards and labelling programme for manufacturers of electrical appliances has been started and is expected to lead to savings of 11,689 million kW annually in the first 5 years of its operation.[35] *Green rating of buildings* is covered by several schemes including the Indian chapter of the Leadership in Energy and Environmental Design (LEED)[36] and the

[34] The Ministry of New and Renewable Energy, which began as a Department of Non-Conventional Energy Sources within the Ministry of Energy in the 1970s was recast as a new ministry in 2006.

[35] Mork, Knut Anton, "Energy Pricing In India :Perspectives Issues, and Options, The Energy Journal, July1989.

[36] Green Building Rating System, developed by the U.S. Green Building Council (USGBC), provides a suite of standards for environmentally

Contd....

indigenous systems, namely green rating for integrated habitat assessment (GRIHA).[37] National Urban Transport Policy emphasizes extensive public transport facilities and non-motorized modes over personal vehicles.[38] The Delhi MRTS expansion and other mass transit systems, such as the Metro Bus project in Bangalore, are steps in its implementation. The policy also favours cleaner fuels like CNG and encourages R&D for commercialization of cleaner technologies.[39] Directly or indirectly these efforts are made by Government as well as by people to reduce energy consumption. Some of these efforts are on-going for several decades and are institutionalized in a number of ways through policies, programmes and the creation of specific institutions.

NATIONAL ENVIRONMENT POLICY

It lays down the overall strategy and programme for the conservation and preservation of environment. It emphasis on certain principles like Conservation of Critical Environmental Resources, Intra-generational and Inter-generational Equity, Integration of Environmental Concerns in Economic and Social Development, Environmental Governance and Enhancement of Resources for Environmental Conservation.[40]

Contd. from previous page...

sustainable construction. Since its inception in 1998, LEED has grown to encompass more than 14,000 projects in 50 U.S. States and 30 countries.

[37] The Indian version of Green Building is developed by TERI. TERI's green building rating evaluates the environmental performance of a building holistically over its entire life cycle, thereby providing a definitive standard for what constitutes a 'green building'. The rating system, based on accepted energy and environmental principles, will seek to strike a balance between the established practices and emerging concepts, both national and international.

[38] National Urban Transport Poliy,2005.

[39] Shreekant Gupta, "Urban Transport in India: Beyond the "Nano",Institute of South Asian Studies Brief, (2008).

[40] M.S. Bhatt, Shahid Ashraf and Asheref Illiyan, "Problems and Prospects of Environment Policy -- Indian Perspective", New Delhi, Aakar, (2008).

These principles directly or indirectly have a clear bearing on climate change. National environment policy also reiterates the broad principle of India's response to climate change which includes:

(*a*) Adherence to the principle of common but differentiated responsibilities and respective capabilities of different countries in respect of both mitigation of GHGs, and adaptation measures.

(*b*) Reliance on multilateral approaches, as opposed to bilateral or plurilateral or unilateral measures.

(*c*) Equal per-capita entitlements of global environmental resources to all countries.

(*d*) Over-riding priority of the right to development.

(*e*) Identify key vulnerabilities of India to climate change, in particular impacts on water resources, forests, coastal areas, agriculture, and health.

(*f*) Assess the need for adaptation to future climate change, and the scope for incorporating these in relevant programmes, including watershed management, coastal zone planning and regulation, forestry management, agricultural technologies and practices, and health programmes.

(*g*) Encourage Indian Industry to participate in the Clean Development Mechanism (CDM) through capacity building for identifying and preparing CDM projects, including in the financial sector.

(*h*) Participate in voluntary partnerships with other countries both developed and developing, to address the challenges of sustainable development and climate change, consistent with the provisions of the UN Framework Convention on Climate Change.[41] An

[41] National Environment policy, 2006.

example is India's membership of the Asia-Pacific Partnership for Clean Development and Climate 2005.[42] Under this partnership Foreign, Environment and Energy Ministers from partner countries agreed to co-operate on development and transfer of technology, which enables reduction of greenhouse gas emissions. In this ministers agreed to a Charter, Communique and Work Plan that outline a "private-public task forces to address climate change, energy security and air pollution."[43] An analysis of India's climate policy response and stands at various international forums clearly indicate the adoption and of these principles.

POLLUTION CONTROL

India's response to the problem of pollution has been in the form of multi-pronged strategies in the form of legislations, agreements, fiscal incentives and other measures. Seventeen categories of heavily polluting industries have been identified under pollution Acts.[44] Submission of an Environmental Statement by polluting units seeking consent either under the Water (Prevention and Control of Pollution) Act, 1974 or the Air (Prevention and Control of Pollution) Act, 1981 or both and the Authorization under the Hazardous Wastes (Management and Handling) Rules, 1989 has been

---

[42] This is an international non-treaty agreement among Australia, Canada, India, Japan, the People's Republic of China, South Korea, and the United States to co-operate on development and transfer of technology which enables reduction of greenhouse gas emissions

[43] Communiqué, Asia-Pacific Partnershipon Clean Development and Climate Inaugural Ministerial Meeting Sydney, 11-12 January 2006 available at http://www.asiapacificpartnership.org/pdf/resources/Communique.pdf (Last accessed on 23-02-2009)

[44] The Industries Are cement, thermal power plant, distilleries, sugar, fertiliser, integrated iron and steel, oil refineries, pulp and paper, petrochemicals, pesticides, tanneries, basic drugs and pharmaceuticals, dye and dye intermediates, caustic soda, zinc smelter, copper smelter and aluminium smelter

made mandatory under the Environment (Protection) Act, 1986. These measures are intended to have some bearing on climate change. The government has not taken any step to reduce carbon dioxide emissions under its pollution control law, nor are any plans to do so imminent. If experience is any indication, the government may not undertake such measures voluntarily--critical pollution controls in the past have been motivated by judicial intervention.[45]

OZONE DEPLETION MITIGATION

Ozone Cell, a special directorate established within the Ministry of Environments Forests (MOEF), dedicated to managing and coordinating the implementation of the Montreal Protocol in India the Ozone Depleting Substances (Regulation and Control) Rules, 2000 was passed.[46] The Ministry provides custom/excise duty exemption for ODS phase-out projects and grants duty exemption for new investments with non-ODS technologies. The Reserve Bank of India has issued guidelines to all financial institutions and commercial banks not to fund new ventures with ODS technology.[47]

ENVIRONMENT IMPACT ASSESSMENT

India introduced the concept of EIA in 1994 under the provisions of Environment (Protection) Act, 1986 making EIA mandatory for 29 categories of developmental projects. [48] The

---

[45] For instance, the government passed legislation adopting the Euro Emissions II standards for all road vehicles, following litigation before the Supreme Court of India. See M.C. Mehta v. Union of India and Others, WP 13020/1985 (1999.04.06).

[46] Ozone Depleting Substances (Regulation and Control) Rules, 2000.Ministry of Environment and Forests, India.

[47] H.A.C. Prasad, Climate Change and India- Some Major Issues and Policy Implications,working papers, Department of Economic Affairs Ministry of Finance Government of India,March, 2009.

[48] Environment Impact Assessment 1994, Impact Assessment Division Ministry of Environment and Forests, Government of India.

programme of EIA was initiated with the appraisal of River Valley Projects EIA now applies to 32 categories of development activities including sectors like industry, thermal power, hydroelectric, nuclear mining, construction projects and infrastructure. Checklist for ecological assessment under Annexure 5 contains various criteria's which has a bearing on climate change.[49]

CLIMATE CHANGE POLICY

India's climate change policy and law primarily flow from global arrangements. India signed and ratified the United Nations Framework Convention on Climate Change (UNFCCC). 10 June 1992 and ratified it on 1 November 1993.[50] The ultimate goal of the UNFCCC convention is to achieve, stable greenhouse gas concentrations in the atmosphere. The convention obliges countries to formulate, publish, and regularly update national and where appropriate regional programmes continuing measures to mitigate climate change, co-operate in preparing for adaptation to the impacts of climate change, develop and elaborate appropriate and integrated plans for coastal zone management, water resources and agriculture and for the protection and rehabilitation of areas affected by drought, desertification and floods.[51] In addition, India has submitted its first National Communication on Climate Change (NATCOM I) and is currently working on NATCOM II.[52]

---

[49] Annexure 5 EIA Notification1994.

[50] United Nations Framework Convention on Climate Change, May 9, 1992, S. Treaty Doc. No. 102-38 (1992), 1771 U.N.T.S. 107

[51] The United Nations Framework Convention on Climate Change (UNFCCC or FCCC) is an international environmental treaty produced at the United Nations Conference on Environment and Development (UNCED), informally known as the Earth Summit, held in Rio de Janeiro in 1992. The treaty is aimed at stabilizing greenhouse gas concentrations in the atmosphere at a level that would prevent dangerous anthropogenic interference with the climate system.

[52] *Supra* note1

The Kyoto Protocol to the UNFCCC was adopted in 1997 and India acceded to the Kyoto Protocol, which is hailed as a major mile stone in climate change mitigation strategy in 2005. But, because of low *per capita* emissions, it enjoys a surplus of emission rights. In other words, unlike Annex I countries such as Australia, Japan, Canada, and Europe, India is not legally required to reduce its emissions within a specific target date. Kyoto Protocol envisages various mechanisms like clean development mechanisms, Joint implementation scheme etc. to reduce GHG emissions. [53]

In pursuance of this, India the administration has established a National Clean Development Mechanism Authority (NCDMA) within the central Ministry of Environment and Forests (MoEF) to coordinate CDM projects.[54] NCDMA has developed procedures for granting approval to projects proposals and it supervises the implementation of the projects. However, poor infrastructure, market viability, and intellectual property regimes that inhibit transfer of low-carbon technologies into India limit the efficacy of CDM undertakings.

Further, the Expert Committee on Impact of Climate Change set up by the Ministry of Environment & Forests in June 2007 assessed the impact of climate change on six areas, namely Water Resources, Agriculture, Natural Eco-system, Health, Coastal Zone Management and Climate Modeling. Reports of the Expert Committee in these areas have been prepared. [55] Specific policies and legislation to mitigate or adapt to climate change were, however, absent. A special committee, the National Council on Climate Change (the

---

[53] *See* Article 12, Kyoto Protocol to the United Nations Convention on Climate Change, Dec. 10 1997, 37 Int'l Legal Materials 32, *available at* http://unfccc.int/resource/docs/convkp/kpeng.pdf

[54] National CDM Authority, CDM India: Designated National Authority, *available at* http://cdmindia.nic.in/cdm_india.htm

[55] Press Release, Ministry of Environment and Forests, Wednesday, October 22, 2008.

Council) established by the Prime Minister of India, led finally to the adoption of National Action Plan on Climate Change. [56]

NATIONAL ACTION PLAN ON CLIMATE CHANGE: SUBSTANTIVE PROVISIONS

India's National Action Plan on Climate Change aims at adopting a low carbon energy path for achieving sustainable development without compromising on economic growth and quality of life.[57] The plan adopts multi-pronged, long-term and integrated strategies for achieving various goals in the context of climate change. NAPCC aims to strike a balance between the need to maintain a high growth rate for increasing the living standards of people and the need to reduce vulnerability of the masses to the impacts of climate change.[58] The operational thrust of NAPCC is in the shape of eight Missions.

National Solar Mission: The National Mission of Solar Energy occupies a pre-eminent place, whose success, according to the Prime Minister, has the potential of transforming the face of India.[59] It aims to promote the development and use of solar energy for power generation and other uses with the ultimate objective of making solar competitive with fossil-based energy options.[60] The stated objective of the mission is to increase the share of solar energy and other renewable and non-fossil based energy sources in the total energy mix of the country. This is justified by the fact that India is ideally situated in the equatorial solar Belt receiving profuse solar radiation throughout the year. The average solar insolation incident over India is about 5.5 kWh/m^2 per day, which means that

[56] National Action Plan On Climate Change, India (2008).

[57] *Id*,Principles, NAPCC.

[58] *Id*

[59] P.Ms speech at the release on National Action Plan on Climate Change,2008.

[60] *Id*, National Action Plan on Climate Change, p. 18.

just 1% of India's land can meet the country's entire electricity requirement till 2030.[61]The plan includes:

- Specific goals for increasing use of solar thermal technologies in urban areas, industry, and commercial establishments;
- A goal of increasing production of photovoltaics to 1000 MW/year; and
- A goal of deploying at least 1000 MW of solar thermal power generation.

Other objectives include the establishment of a solar research center, increased International collaboration on technology development, Strengthening of domestic manufacturing capacity, and Increased government funding and International support. [62]

National Mission for Enhanced Energy Efficiency: Energy efficiency has always been on the frontal agenda of India's environment policy, Various Acts has been passed in order to achieve the above objectives including the Energy Conservation Act 2001.[63] Initiatives under the plan are expected to yield savings of 10,000 MW by 2012. The plan recommends:

- Mandating specific GHG mitigation options in large energy-consuming industries, with a system for companies to trade energy-savings certificates;
- Encouraging financing for public-private partnerships to reduce energy consumption through demand-side management programmes in the municipal, buildings and agricultural sectors and provide them with cross cutting technology options; and

---

61 *Id*

62 *Id*

63 Energy Conservation Act 2001, provide for efficient use of energy and its conservation.

- Accelerate the shift to energy efficient appliances in designated sectors and provide them with schemes in the form of reduced taxes.[64]

Further the plan focuses on tradable certification of energy efficiency, develop fiscal instruments to promote energy efficiency and to make energy efficient products affordable.

National Mission has already proposed an Energy Efficiency Action Plan which emphasizes Perform Achieve and Trade scheme, Market Transformation for Energy Efficiency (MTEE) in the form of CDM roadmap, Standards and Labeling, ESCO promotion, capacity-building, Financing Energy Efficiency through Tax exemptions, Revolving fund, Partial Risk Guarantee Fund.[65] Currently this action plan is open for public comment.

National Mission on Sustainable Habitat: To promote energy efficiency as a core component of urban planning, the plan calls for:

- Extending the existing Energy Conservation Building Code;
- A greater emphasis on urban waste management and recycling, including power production from waste; and
- Stronger enforcement of automotive fuel economy standards, using pricing measures to encourage the purchase of efficient vehicles, and providing incentives for the use of public transportation.[66]

National Water Mission: Climate change is expected to adversely affect the water supply of India. Accordingly the plan sets a goal of a 20% improvement in water use efficiency through pricing and other measures.

---

[64] *Supra* note 56 See NAPCC.

[65] Namrata Kala, Energy Efficiency In India overview, and Future Out look, TERI.

[66] *Supra* note 56 NAPCC.

National Mission for Sustaining the Himalayan Ecosystem: Himalaya houses an ecosystem unique for its diversity. The plan aims to conserve biodiversity, forest cover, and other ecological values in the Himalayan region, where glaciers that are a major source of India's water supply are projected to recede as a result of global warming. It also reaffirms the following measures mentioned in the National Environment Policy, 2006.

- Adopting appropriate land-use planning and watershed management practices for sustainable development of mountain ecosystems.
- Adopting best practices for infrastructure construction in mountain regions to avoid or minimize damage to sensitive ecosystems and despoiling of landscapes.
- Encouraging cultivation of traditional varieties of crops and horticulture by promoting organic farming, enabling farmers to realise a price premium.
- Promoting sustainable tourism based on best practices and multi-stakeholder partnerships to enable local communities to gain better livelihoods.
- Taking measures to regulate tourist inflows into mountain regions to ensure that the carrying capacity of the mountain ecosystem is not breached.
- Developing protection strategies for certain mountain scopes with unique "incomparable values".[67]

National Mission for a "Green India": Goals include afforestation of 6 million hectares of degraded forest lands and expanding forest cover and density cover from 23% to 33% of India's territory.

National Mission for Sustainable Agriculture: Agriculture will be one of the critically affected sectors of climate change.

---

[67] National Environment Policy, 2006.

The plan seeks to carry on climate adaptation in agriculture through various scientific agricultural practices like dry land agriculture, the advancement of climate-resilient crops. In addition, the Mission makes suggestions for safeguarding farmers against increased risk due to climate change. These suggestions include, strengthening agricultural and weather insurance; creation of web-enabled, regional language based services for facilitation of weather-based insurance; development of GIS and remote sensing methodologies; mapping vulnerable regions and disease hotspots; and developing and implementing region-specific, vulnerability based contingency plans.[68] Finally, it suggests greater access to information and use of biotechnology.[69]

National Mission on Strategic Knowledge for Climate Change: To gain a better understanding of climate science, impacts and challenges, the plan envisions a new Climate Science Research Fund, improved climate modeling, and increased international collaboration. It also encourages private sector initiatives to develop adaptation and mitigation technologies through venture capital funds.[70]

The NAPCC also describes certain restrictions and conditions on the existing ongoing programmes For instance the government is contemplating the makeover of inefficient coal-fired power plants and supporting the research and development of IGCC and supercritical technologies. The plan lays down that under the Electricity Act 2003 and the National Tariff Policy 2006, the central and the state electricity regulatory commissions must purchase a certain percentage of grid-based power from renewable sources. Further under the

---

[68] *Focus: National Action Plan on Climate Change, Corporate Planning and economics Studies Department, Vol 31, 10th December 2008.*

[69] India's National Action Plan on Climate Change, available at http : / / www.climate-leaders.org / climate-change-resources / india-and-climate-change / indias-national-action-plan-on-climate-change#foot_1(last accessed on 3-03-2009.

[70] *Supra* note 56.

Energy Conservation Act 2001, large energy consuming industries are required to undertake energy audits and an energy labeling programme for appliances has been introduced.

The implementation of the plan has been left to the lead ministries. They are directed to develop objectives, implementation strategies, timelines, and monitoring and evaluation criteria, to be submitted to the Prime Minister's Council on Climate Change. The Council will periodically review and report on each mission's progress. [71]

NAPCC: Milestone in India's Climate Change Policy

The NAPCC has laid down a national strategy on climate change with regard to adaptation and mitigation of greenhouse gas emissions, both domestically and in the context of global arrangements under the United Nations Framework Convention on Climate Change (UNFCCC). Sustainable development is the key theme running the breadth of the plan. The action plan does not set target reduction of greenhouse gas emissions but seeks to promote sustainable development through use of clean technologies. India is a big player in climate change discourse and India's stand can provide direction to climate change strategy especially for developing countries. The plan will definitely give a boost to India's suitability at the global level to augment our bargaining control with the developed nations, especially for providing clean technologies at cheaper rates. For the first time the plan chalks out an *institutional structure* for future action overseen by the Council on Climate Change headed by Prime minister.[72]

The Action plan in various ways enunciates a new approach different from the earlier existing climate change mitigation strategies. Firstly adaptation to the adverse effects

---

[71] *Id*

[72] *Id*

of climate change is considered the policy priority under the plan. This is based on the Intergovernmental Panel on Climate Change (IPCC) predictions that the adverse effects on climate change will have implications for food security, livelihood health etc. whereas international and even our earlier approach were mainly concerned with energy (oil) security even though ensuring food security is a key objective of the climate change convention.[73]

Secondly, it stresses qualitative shifts rather than quantitative reductions of emissions which is the high point attracting the west. Developed countries have not even been able to agree on the base year to measure reductions, the percentage reductions to be brought about and the quantities of carbon that can be eliminated without affecting economic growth because each country looks at the problem through the prism of its national industry.[74] Our plan makes a positive shift from this quantitative game strategy. Thirdly the stress of the plan is on modifying activities that generate emissions of carbon or are affected by climate change, through new knowledge. The key role of technology establishes a bottom-up approach and a basis for international cooperation very different to the current stress of developed countries on a global agreement for determining when worldwide emissions should peak.[75]

Critics have observed that the country 'faces an uphill task of not only convincing the global community but even environmentalists in the country about the feasibility of its action plan.[76] Though it is commonly agreed that the NAPCC represents a significant step forward, in that it is the first systematic attempt by the Government to frame a comprehensive policy framework to deal with climate change,

---

[73] Mukul Sanwal, The G-8 and India's National Action Plan on Climate Change, Economic & Political Weekly, (2008).

[74] *Id*

[75] *Id*

[76] P.P. Sangal, India's climate change action plan, *The Economic Times*, Mumbai, July 29, 2008.

most believe that it is insufficient and real measurable targets. If we analyse the plan we do not commit to specific emission reduction targets or energy efficiency targets. Instead it states that India's per capita GHGs emission would at no point exceed that of the developed countries.[77] NACPP reiterates India's adherence to the Kyoto principle of "common but differentiated responsibilities.[78] The underlying principle behind this seems to be the fact that India's per capita emission is just above one tonne compared to their average per capita emission of other nations. But this will attract flak from western developed nations. Though India's stand is motivated by its developmental needs like tackling poverty and giving boost to industrialisation. The stand does not go with India's image of a responsible player and nation in world affairs.

The greater shortcoming is the failure of the NAPCC to articulate a proper vision, nationally or globally. While espousing a qualitative shift towards ecologically sustainable growth, the plan fails to develop, a vision of sustainable development.[79] The plan does not specify well defined goals and targets to be achieved under the missions in a time bound manner. The "directional shift in the development pathway" that will not compromise on the need for growth while trying to achieve ecological sustainability has been presented as the core of the action plan. However, the action plan fails to explain effectively the nature and direction of that shift.

Various other short comings are pointed out. NAPCC in a way merely combines the Government's existing National Plans for water, agriculture, renewable energy, energy efficiency, etc, rather than formulating a new "well-debated and thought out 'strategy' chalking out a discernible low-carbon pathway for India." Further projections made out in NAPCC are based on the IPCC predictions. This means that

---

[77] *Id*

[78] *Supra* note 56

[79] Climate Change: Not Vision, Not Plan, Economic & Political Weekly, p 5, July 12, (2008).

best available climate science predictions are not taken into account lacking objectivity and scientfity. For instance the NAPCC suggests glacier melt might not be caused by climate change and that further study is needed.[80] Further first paragraph of the plan regarding *Overview* directly blames anthropogenic factors and developed countries for creating the threat of climate change. Even the IPCC has not so vehemently taken the view that human kind is solely responsible for causing climate change.

Even the eight missions which are the focus area under the plan suffer from certain serious short comings. For instance at present, most of our energy to be specific nearly 80% of it comes from burning of fossil fuels which is a major source of GHGs. The action plan gives emphasis to production of solar energy and aims to increase its share. The NAPCC's Solar Mission promises to promote use of solar energy in homes, commercial establishments and for power generation through photo-voltaics, solar thermal and Concentrating Solar Power.[81] There is no target set for increasing the proportion of renewable energy in the total energy mix. But what is to be noted that the identification of nuclear energy is missing from the plan this is quite noticeable in the wake of our nuclear deal with the U.S government. Another area neglected is wind energy which also has tremendous potential and its scope has been assessed at 45,000 MW.[82] In the water sector, the plan attempts to push for more big dams, irrigation projects, hydropower projects, interlinking of rivers and such other long-distance water transfer projects. "This was despite the increasing evidence of non-performance of such projects.[83] Further regarding Municipal solid waste management the

---

80 *Supra* note 56

81 Raghu, "India's Climate Action Plan: Many Points, No Direction" ,http://www.politicalaffairs.net/article/view/7100/1/342/

82 Supra note 69

83 Gargi Parsai, Action plan on climate change lacks transparency, The Hindu, March 2, 2009.

plan takes pride in mentioning that municipal solid waste (MSW) management should be considered a public good or environmental service to be provided by government and simply forgets its business sense and profitability behind it. Further there is nothing new in the plan of increasing forest cover from 23% to 33% the existing environment policy already contains provisions for it. The plan merely reiterates the existing mechanisms.

Other criticisms mainly concerned with the actual implementation and reach of the plan. The question arises will it remain and die as a mere declaration of principles and goals to achieve. Various questions concerning interpretation, and translation of the document and its objectives? Another pertinent issue relates to How is the NAPCC to reach citizens all over India? Who will fund the national outreach/public awareness programmes? An analysis of the framing of the plan reveals that there has been no stakeholder consultation during the framing of the NAPCC. No consultation and opinion seeking was undertaken from a broader community of experts, citizen's groups, civil society, etc. NAPCC does not adequately analyse regional or global views. Specifically NAPCC does not address the issue of deadlock in the international arena on climate change.[84] The Government has not clearly stated how it will finance the plan, it does not talk of any concrete financial mechanisms nor does it set out a budgetary allocation process.[85]

The action plan at various places elaborates and deals with the issue of poverty and emphasized on the need to address their removal. But in actual terms the possible impact of the

---

[84] Rahul Goswami, Blind Spots in India's new National Action Plan on Climate Change; Indian Network on Ethics and Climate Change; http://www.inecc.net/debate/Blind%20Spots%20in%20NAPCC.php(last accessed on 1-02-2009).

[85] Tirthankar Mandal; Climate Brief 5: The Action Plan on Climate Change, G8 Declaration and the Accra Climate Change Meet: Points to Ponder; Centre for Trade and Development; August 2008; http://www.centad.org/download/Climate_Brief_5.pdf(last accessed on 10-02-2009).

plan on people below poverty line and remote areas is questionable. On the contrary, the implementation of the Plan could impact them adversely if the economic burden of the adaptation strategies is not supported by the government. The majority of the vulnerable population of India is poorly equipped to cope effectively with the adversities of climate change due to low capabilities, weak institutional mechanisms, and lack of access to adequate resources.[86] The role of government is recognized and talked about only as a regulator and a facilitator and as a consumer is not acknowledged; this is quite perplexing given the fact that the government is the largest consumer in India. Further Plan envisages eight national missions if past is of any lesson, Coordination between different missions with elaborate structures of cross-cutting activities will be a problem.

The problem of intellectual property rights (IPRs) in connection with technology transfer from developed countries is clearly missing from the focus of the plan. Unfortunately plan nowhere recognises that the issue of IPRs in climate-related technologies is fundamental to the North-South conflict as well as to tackling the climate crisis, India should in fact follow la principle and demand from the west and international forums that all climate-related technologies should be placed in the public domain rather than allowing certain corporations or nations to take profits on an international issue which is mainly created by their blind pursuit of growth.[87]

CONCLUSION

Climate change has come to be regarded as one of the key global challenges of our time and India is no exception. In policy terms, it is no longer thought of as an environmental issue only, but as a challenge with potentially far-reaching

---

86 *Supra* note 9

87 *Id*

implications for virtually all areas of planning, from health care, to transport, to taxation. As the fifth largest emitter of carbon dioxide, India's position and engagement is critical to the future of an effective international climate regime, particularly when one considers the potential effects that a rapidly growing hydrocarbon economy of a billion plus people can have on the global emissions, and hence the climate. This is a legitimate concern and one that requires attention; and indeed, the Indian government has been responsive and is presently ahead in the curve of mitigation trends.

What should be remembered is that however, India's efforts to mitigate climate change, or for that matter taking an international position challenging industrialized countries to reduce emissions, cannot ensure that fundamental rights of millions of Indians will be protected if there is a two or more degrees Celsius increase in global temperature. Catastrophic events that may be unraveled by climate change will not only have severe economic repercussions, but also implicate constitutional rights of citizens. The rights that will be affected are either fundamental rights under the Indian Constitution, including right to life, or rights that the Supreme Court of India has held to be an integral part of this right to life, including rights to livelihood, health and basic necessities.

Thus in India, the major concern is securing compliance to these policies. Climate change provides a unique opportunity to make India's development path people and environment-friendly. The costs of not addressing climate change or to adapt to it are very uncertain, but their welfare consequences are enormous. Early actions on adaptation therefore are prudent and consistent from the viewpoint of 'precautionary principle'.[88]

---

88 Jayant Sathaye, P. R. Shukla and N. H. Ravindranath, "Climate Change, Sustainable Development and India: Global and National Concerns", *CURRENT SCIENCE*, VOL. 90, NO.3, p 314, 10 FEBRUARY (2006).

3

Changing Dimensions of Renewable Energy Policy in Response to Climate Change

INTRODUCTION

Climate change occurs as a consequence of anthropogenic activities like combustion of fossil fuels, industrial processes, deforestation and GHGs release into atmosphere. Power sector alone contributes to the scale of 40% to the total carbon emissions.[1] Climate change affects populations by changing basic life conditions, for example, food availability, and by causing habitat loss and fragmentation.[2] The global concern for sustainable development and climate change has brought attention of policy makers towards the renewable energy sources since these provide energy, without emissions of greenhouse gases (GHGs) and are also abundant resource available for future. This chapter analyses the existing renewable energy policies in India and throws light on the reasons for failure of renewable energy projects as commercial ventures irrespective of their huge potential in India. Chapter proposes policy instruments that determine the style and cost-

[1] Shukla P.R., Ghosh D., Chandler W., Logan J. "Developing countries and Global Climate Change: Electric Power Options in India". Prepared for the Pew Centre on Global Climate Change, Arligton, US,(1999).

[2] Crick, H.Q.P., Sparks, T.H." Climate change related to egg laying trends". *Nature* 399, 423, (1999).

effectiveness of renewable energy projects implementation and create a framework for financing mechanisms for these projects. Suggestions have been made to reduce the dependence on energy based on fossil fuels. Pollution prevention and dependency on the fossil fuels can be achieved by government's directly regulating pollution-generating activities or, indirectly, by influencing the decision making process on the micro level (economic instruments). Chapter recommends a sustainable energy policy that will not only meet the future energy demand for rapid economic growth but also protect the environment and conserve scarce resources.

INCREASING ENERGY DEMAND AND NEED FOR RENEWABLE ENERGY

Energy is the sine qua non of development. India, with over a billion people, today only produces 660 billion KWh of electricity and over 600 million Indians, a population equal to the combined population of USA and EU, have no access to electricity, and limited access to other clean, modern fuels such as LPG and kerosene.[3] Besides this shortage of power adversely impact our industry and other development process. With the growing population and development process energy demand in India is expected to rise (Table 3·0) Augmenting energy production and supply becomes key to India's development. India's future energy requirements are going to be very high and renewable energy can be one of the efficient and eco-friendly ways to meet the same.

ENERGY DEMAND PROJECTIONS

An analysis of India's energy generation and consumption brings that India, with a major reserve of coal, depends on an

[3] "India :Addressing Energy security and Climate Change", Ministry of Environment & Forests, Ministry of Power Bureau of Energy Efficiency Government of India (2007).

energy system that is exceedingly carbon centric. The carbon centric energy profile suffers from certain inherent drawbacks and makes a case for shifting to renewable energy sources.

Table 3.0: Energy Demand Projections

Year	*Mtoe*
2011-12	677
2016-17	861
2021-22	1,082
2026-27	1,417
2031-32	1,818

Source: Energy Policy Report, Planning Commisssion, India

Firstly these non renewable carbon based source are clearly depletable meaning one day our energy supply will not be able to cope up with the increasing energy demand. Current estimates show total oil reserves of 5.7 billion barrels, enough to last a dozen years or so at the present levels of consumption. Recent international studies indicate that no more than a quarter of India's twenty-six secondary basins have seen serious exploration activity. These include offshore basins covering about 3,80,000 square kilometres and on shore basins covering 13,40,000 square kilometres. It has been estimated that these untapped basins may well contain thirty billion tonnes of reserves, five times the current figure, thereby extending the projected depletion date by more than fifty years[4] of consumption, at 0.795 trillion cubic feet already exceeded production in 2000.[5] These are already evident from the power shortage faced by the country. India's current electricity installed capacity is 135 401.63MW. Currently there is peak power shortage of about 10 % and overall power

[4] 'Hydrocarbons Vision 2025', Directorate General of Hydrocarbons, Government of India.

[5] An Energy Overview of India', EIA, USA, available at http://www.fe.doe.gov/international/indiover.html (last accessed on 1-04-2009)

shortage of 7.5 %. Around 45 % household are still out of basic electricity services. Clean cooking energy is a major issue. Dependency on imported oil is increasing. Besides, large quantities of traditional biomass resources used for the energy demands of the vast rural population are exerting pressures on forests and village woodlots. Secondly there is an environment perspective to it. Experts believe that major cause of climate change is the release of green house gases due to burning of fossil fuels. Reports suggest that India is the world's fourth largest economy and fifth largest greenhouse gas (GHG) emitter, accounting for about 5% of global emissions.[6] India's emissions increased 65% between 1990 and 2005 and are projected to grow another 70% by 2020.[7] In order to make matters worse India's first National Communication (2004) reveals that the energy sector accounts for around 61 per cent of total national emissions. For fossil fuels, coal combustion had a dominant share of emissions, amounting to around 64 per cent of all energy emissions.[8] Thus, the energy scenario in India is clearly unsustainable.[9] This fact probably forced Indian prime minister to make the following statement. "Our vision is to make India's economic development energy-

---

[6] Climate Change Mitigation Measures in India, International Brief, prepared by Global Pew Climate Center, September 2008.

[7] Data from IEA, 2007. "CO_2 Emissions from Fossil Fuel Combustion 1971-2005"; IEA, 2007. "World Energy Outlook 2007: "Global Anthropogenic Non-CO_2 Greenhouse Gas Emissions: 1990-2020". All emissions figures are for total greenhouse gas emissions as estimated by the International Energy Agency (CO_2 from fossil fuels), the US EPA (other greenhouse gases) and the World Energy Outlook (projections), except for GHG emissions by sector which rely on these same sources but are estimated by Ecofys, 2006 in this report: available at http://www.ecofys.com/com/publications/documents /Ecofys_Factor s_underpinning_future_actionOct2006.pdf. (last accessed on 1-04-2009).

[8] India's National Communications to the United nations Frame Work Convention on Climate Change,Ministry of Environment and Forests, New Delhi, 2004.

[9] Loulou R, Shukla P R, and Kanudia, A.," Energy and Environment Strategies for a Sustainable Future: Analysis with the Indian MARKAL Model", Allied Publishers, New Delhi,(1997).

efficient. Over a period of time, we must pioneer a graduated shift from economic activity based on fossil fuels to one based on non-fossil fuels."[10] This Announcement of the prime minister, unveiling the national action plan on climate change clearly indicates the importance India attaches to renewable energy policy.

RENEWABLE ENERGY SOURCES

India is favourably situated with regard to various renewable energy sources and has an abundance of sunlight, water and biomass. Vigorous efforts during the past two decades are showing result as the emphasis is now slowly shifting towards renewable energy. In fact India has the world's largest programme for renewable energy.

Wind Power

India now ranks as a "wind superpower" with an installed wind power capacity of 1167 MW and about 5 billion units of electricity have been fed to the national grid so far.[11] In progress are wind resource assessment programme, wind monitoring, wind mapping, covering 800 stations in 24 states with 193 wind monitoring stations in operations. Altogether 13 states of India have a net potential of about 45000 MW. Wind power in fact constitutes over 65% of renewable capacity, ranking India fourth in terms of wind power generation worldwide.[12] India's wind energy has grown dramatically since its first wind farm was set up in Mandvi in the state of Gujarat in 1986 with a capacity of just 55 MW. It has now just surpassed Denmark, one of the pioneers of the technology, and provides sixty per cent of the electricity the country derives from renewable.

Further this is one of the fastest growing sectors in the country.

---

[10] National Action Plan On Climate Change, India, 2008.

[11] Renewable Energy In India :A Purview, Status of renewable energy technologies in India,p28.

[12] "Annual Report 2006- 07". Government of India, Ministry of New and Renewable Energy, 2007.

At the rate at which new wind energy farms are being set up in India, its capacity from the wind will exceed 5,000 MW in less than two years, Geographically the most favourable location in case of wind power is the southern India specially Tamilnadu as it gets winds from the South West and the North East. As a result, a 1 MW machine there can generate 3.5 million units a year as compared to 2 million units in a western state like Maharashtra. In fact More than three fifths of India's wind power is generated in the southern state of Tamil Nadu.

Solar Power

Because of its location between the Tropic of Cancer and the Equator, India has an average annual temperature that ranges from 25°C - 27.5 °C. This means that India has huge solar potential. The sunniest parts are situated in the south/east coast, from Calcutta to Madras.

Solar energy is inexhaustible. It is also widely distributed, environment friendly and cost free in raw form. Partially offsetting these manifold benefits are two technological disadvantages: Its low intensity or dispersion and its unpredictability, which varies with the whims of weather.[13] Solar energy has several applications: photovoltaic (PV) cells are placed on the roof top of houses or commercial buildings, and collectors such as mirrors or parabolic dishes that can move and track the sun throughout the day are also used. This mechanism is being used for concentrated lighting in buildings. Photovoltaic (PV) cells have a low efficiency factor, yet power generation systems using photovoltaic materials have the advantage of having no moving parts. PV cells find applications in individual home rooftop systems, community street lights, community water pumping, and areas where the terrain makes it difficult to access the power grid. The efficiency of solar photovoltaic cells with single crystal silicon

---

[13] *Shikha Bisht & Biswajayee Patra,* Solar Energy: Alternative to Combat Energy Insecurity in India, Article No:96, December 12, 2006.

is about 13 % to 17%. High efficiency cells with concentrators are being manufactured which can operate with low sunlight intensities.[14] India has the world's largest deployment of solar PV (Photovoltaic) consisting of 850,000 PV systems aggregating 50 MW. India ranks 3rd in annual production capacity of solar PV (17 MW) ahead of Germany, France and Australia.[15]

Hydro power

Hydropower is a renewable economic, non-polluting and environmentally benign source of energy. India has immense economically exploitable hydro power potential. It ranks fifth in the world in terms of usable potential. This is distributed across six major river systems. The total potential has been assessed to 84,000MW at 60% load factor. (14870MW Installed capacity). Out of the total capacity of 30,164 MW (20.3%) has been developed so far and 13,616 MW (9.2 %) of capacity is under construction (Table 3·1). There exists huge gap between potential and the actual installed project.

Table 3·1: Area -Wise Potential and its Current Position of Progress at 60% Load Factor as on 1 January 2005

Region	*Potential Assesssed MW*	*Potential Installed (MW)*	*Potential Under development (MW)*
Northern	30,155	5,150	2,905
Western	5,679	2,270	1,164
Soutern	10,763	5,924	153
Eastern	5,594	1,364	201
N.Eastern	31,857	517	914
Total	84,044	15,225	5,339

Source: Author Data Indian National Hydro Power Association. 2005.

---

[14] http://www.geni.org/globalenergy/library/energytrends/currentusage/renewable/Renewable-Energy-Potential-for-India.pdf

[15] *Supra* Note 13

In addition, 6,782 MW in terms of installed capacity from small, mini and micro hydro schemes have been assessed. Also, 56 sites for pumped storage schemes with an aggregate installed capacity of 94,000 MW have been identified. The government expects to harness its full potential of hydropower by 2027 with a whopping investment of 5,000 billion Rupees.[16] Regarding India, small hydro is the most utilized and suitable renewable energy source for energy production. Compared to Small and mini hydel projects they have the potential to provide energy in remote higher budgetary allocation for the hydro sector; investment approval of new projects and identification of new projects.

India has huge reserves and potential for developing various Renewable Energy. Renewable energy adoptions are a positive step toward sustaining our economic growth as well as mitigating climate change phenomenon.

STATUS OF RENEWABLE ENERGY PLANTS

India is rich in coal and abundantly endowed with renewable energy in the form of solar, wind, hydro and bio-energy. India has a vast potential of renewable energy sources and a number of technologies have been developed to harness them. Even then nearly 62 per cent of Indian power generation is from coal fired thermal power plants and 70 per cent of the coal produced every year in India has been used for thermal generation (Fig. 3·1).

Out of the available installed capacity of around 1,44,336 MW electricity, 91, 907 MW is contributed by Thermal energy making it the largest share in energy profile. Hydro power contributes about 35, 909 MW and 4, 120 MW of contribution comes form nuclear energy. Surprisingly Renewable energy contribution is limited to a mere 12,400MW.This clearly depict the low level of share of renewable energy to India's power

16 Avilash Roul, Indias Hydro Power:Can India Achieve its Potential available at http://www.ecoworld.com/features/2007/10/17/indias-hydro-power/ (Last accessed on 3-03-2009).

Total Installed Capacity of Power Distribution India (1, 44,336 MW) March 2008

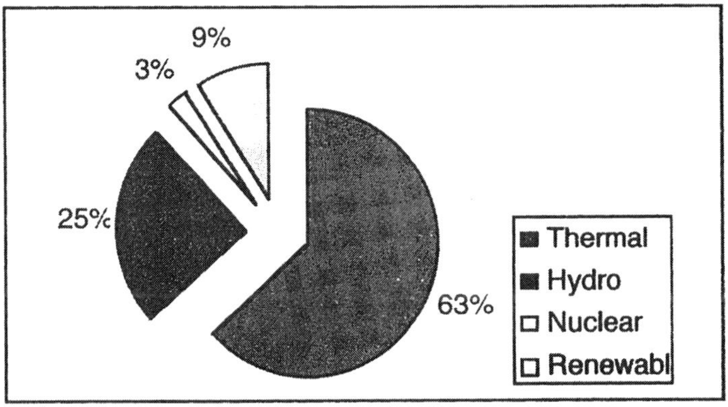

Figure 3·1: India Power Sector Distribution

Source: Author

sector. At the same time various reports clearly indicates that India has a large potential for developing renewable energy. At present there is a huge gap between the potential and the installed capacity and achievement.

In case of centralized renewable energy plants, Wind energy has made significant contributions to India's energy supply. But solar and biomass power plants are yet to make any contribution. Solar is yet to make a start and there is not even a single power plant with capacity more than 1 MW. Even if a fraction of potential of solar (20MW/sq km) is harnessed, it can make significant contribution to energy sector. However solar has made some contribution in decentralized systems mostly in rural areas. Decentralized systems are very useful to the rural areas where there is not supply of electricity and demand is low to set up power lines from grid. For example, Solar PV pumps traditionally have been a huge success in pumping water in remote areas and can be further expanded with technology advancements and incentives from government. It is also noted that the existing gap between potential and achievement is equally distributed among different sources of renewable. (Table 3.2)

Table 3·2: Renewable Energy Potential and Achievement in India[17]

Source/Sector	*Potential*	*Achievement*
A. Power From Renewables		
1. Wind	45,000 MW	2,002 MW
2. Solar Photovoltaic Power	20MW/sq km	47 MW
3. Small hydro (upto 25 MW)	15,000 MW	1,520.35 MW
4. Biomass Cogeneration power	19,500MW*	570.9MW
5. Energy Recovery from waste	1700 MW	30.8 MW
Total power from Renewables	**81, 200 MW**	**4127. 37 MW**
B. Decentralised Systems		
1. Solar water heating (Collector area)	140 Million m^2	0.70 Million m^2
2. Solar PV	20 MW/sq.km	47MW
(*a*) Street lighting systems	-	47969Nos.
(*b*) Home lighting systems	-	256673 Nos.
(*c*) Solar lanterns	-	509894 Nos.
(*d*) SPV power Plants	-	1637 kW
3. Solar PV Pumps	-	6368 Nos.
4. Wind pumps	-	945 Nos.
5. Hybrid Systems	-	199.35 kW
6. Biogas Plants	12 Millions	3.44 Millions
7. Improved Chullas (wood stoves)	120 Millions	35.89 Millions
8. CBP/IBP/NBP plants	-	3,902 Nos.

MW= Mega Watt; sq m =Square meter; kW=kilo watt;* Gasifiers are included

Two decades of our experience with renewable energy clearly has resulted in only a slow growth. It is also expected that current growth rate of renewable contribution will

17 New and Renewable Energy Policy Statement 2005 draft report II, Ministry of Non-Conventional Energy Sources (2005), available at http://www.mnes.nic.in/Rene%202005_new.pdf (Last accessed on 30-10-2008).

continue at a slower pace and energy from renewable is not expected to increase its share in the fuel-mix in any significant manner at least up to 2051-52. Under the likely and optimistic scenarios its share is expected to increase to 39.23 per cent.[18]

EVOLUTION OF RENEWABLE ENERGY POLICY IN INDIA

India is one of the few countries to have a major renewable energy programme for nearly two decades, and is the only country to have a full-fledged national ministry to deal with renewable energy. India's Renewable energy was launched primarily as a response to the professed rural energy crisis in the 1970s. Initiated with a target oriented supply push approach, it primarily sought to develop niche applications, such as in rural areas where grid electricity was unavailable. Cash subsidies were and are still being provided for promoting renewable energy technologies (RETs) in India. CASE (Commission on Additional Sources of Energy) was created in 1980, and then the DNES (Department of Non-conventional Energy Sources) was set up in September, 1982.[19] In the initial stages of the programme, the technologies were not mature and there was little international experience in implementation. However, renewables were promoted as a panacea to the energy problems, and doing 'too much too soon' resulted in unrealistic expectations leading to failures.[20] In the early nineties, under the economic liberalization process, the programme received an impetus with a shift in emphasis from purely subsidy-driven dissemination programmes to technology promotion through the commercial route. DNES was converted into a fully-fledged

---

[18] *Id*

[19] Sinha CS. "Renewable energy programmes in India: Some Recent Developments". Natural Resources Forum 18(4):213–24, (1993).

[20] Ramana PV, Shukla PR. "Climate Change Policies and long-term Rural Energy Trends in India. – In: Shukla PR, editor. *Climate Change Mitigation: Shaping the Indian Energy Scenario.* New Delhi, India, Allied Publishers,(1998).

Ministry. Ministry of Non-conventional Energy Sources, or MNES in July 1992, making India the only country in the world with a ministry dedicated to promoting renewable energy technologies (RETs).[21] In 2006 Ministry was renamed as Ministry of New And Renewable Energy. The governance framework in the energy sector is quite complex. Government decision-making on energy at the Central level is distributed between the Ministry of Petroleum and Natural Gas, the Ministry of Coal, the Ministry of New and Renewable Energy (MNRE), the Ministry of Environment and Forests, the Ministry of Atomic Energy, and the Ministry of Power.[22] While the MNRE is responsible for policy creation and promotion of renewable energy programme in the country, the Ministry of Power is responsible for the legal framework Strategy around electricity generation and transmission. The responsibility for comprehensive rural electrification is shared between two different ministries; another fact to be noted here is that Energy being a concurrent subject under the constitution, and the states share powers with the Centre for energy development as well as environmental regulation.

Key elements of renewable energy in existing energy regulations

Several electricity policies in the last few years have talked about the need and priority to promote Renewable energy. The 11th plan target is to add 100 000 MW electricity by 2012 and MNRE has set up target to add 14500 MW by 2012 from new and renewable energy resources out of which 50 MW would be from solar energy.[23] The Indian Energy

---

[21] Annual Report 2000-2001, Ministry of Non-conventional Energy Sources, New Delhi.

[22] Governance of Renewable Energy sector, New Energy India, available at http://www.newenergyindia.org/Governance%20Page.htm (Last accessed on 10-03-2009).

[23] 11th Plan proposals for New and renewable Energy, Ministry of New and renewable Energy, Government of India, (2006).

Conservation Act 2001 provides a policy framework for the development of energy efficiency market. This legislation seeks to implement energy efficiency policies that lead to widespread market development though better standards for appliances and equipment, energy efficiency labeling, rational cost-of-service based tariffs, mandatory energy audits, awareness and training, financial and fiscal incentives.[24] Several electricity policies in the last few years have talked about the need and priority to promote RE. Foremost amongst them is the Electricity Act 2003 which de-licensed standalone generation and distribution systems in rural areas.[25] The Electricity Act, 2003 recognizes the significance and importance of promotion of efficient and environmentally benign policies. Section 3 (1) of the Electricity Act 2003 requires the Central Government to formulate, inter alia, the National Electricity Policy in consultation with Central Electricity Authority (CEA) and State Governments. The section reads. "*The Central Government shall, from time to time, prepare the National Electricity Policy and tariff policy, in consultation with the State Governments and the Authority for development of the power system based on optimal utilization of resources such as coal, natural gas, nuclear substances or materials, hydro and renewable sources of energy". This provision shows the emphasis which the government attaches to Renewable energy policy.*[26]

Further Section 61 (h) of Electricity Act, 2003 provides that the appropriate Electricity Regulatory Commission while specifying the terms and conditions for determination of Tariff shall be inter-alia guided by the need for promoting co-generation and generation of electricity from the renewable sources of energy.[27] The most specific section which lays

---

[24] S. Padmanaban, Triggering Markets for Energy Efficiency in India,p16.

[25] Electricity Act, 2003

[26] *Id*, Section 3(1)

[27] *Id*, Section 61 (h)

down an obligation on the authorities is contained under Section 86 (1) (e) of the Act which provides that one of the functions of the Regulatory Commission is to promote cogeneration and generation of electricity from renewable sources of energy by providing suitable measures for connectivity with the grid and sale of electricity to any person, and also specify, for purchase of electricity from such sources, a percentage of the total consumption of electricity in the area of a distribution licensee.[28]

National Electricity Policy 2005 also emphasize on the development of renewable energy policy and stresses the need for urgent electrification.[29] Clause 5.12.1 of the policy points out that " Non-conventional sources of energy being the most environment friendly there is an urgent need to promote generation of electricity based on such sources of energy. For this purpose, efforts need to be made to reduce the capital cost of projects based on non-conventional and renewable sources of energy. Cost of energy can also be reduced by promoting competition within such projects. At the same time, adequate promotional measures would also have to be taken for development of technologies and a sustained growth of these sources. "The policy provide for optimal utilization of renewable sources of energy and role of State Energy Regulatory Commissions (SERC)'s in promoting cogeneration and non-conventional energy sources.[30]

The National Tariff Policy 2006 envisages certain incentives and steps for the promotion of Renewable energy. It stated that a minimum percentage of energy, as specified by the Regulatory Commission, is to be purchased from renewable sources by April 1, 2006.[31] The policy has following

---

[28] *Id*, Section 86 (1) (e)

[29] National Electricity Policy 2005, PART I - Section 1 published by Ministry of Power, Extraordinary, The Gazette of India,2005.

[30] *Id*, Clause 5.12.2.

[31] National Tariff Policy, Ministry of Power, Government of India 2006.

features. Clause 6.4 of the Tariff Policy observes that, It will take some time before non-conventional technologies can compete with conventional sources in terms of cost of electricity. Therefore, procurement by distribution companies shall be done at preferential tariffs determined by the Appropriate Commission.[32] Policy further lays down that (2) Such procurement by Distribution Licensees for future requirements shall be done, as far as possible, through competitive bidding process under Section 63 of the Act within suppliers offering energy from same type of non-conventional sources. In the long-term, these technologies would need to compete with other sources in terms of full costs. In case procurement is not done through competitive bidding. The Central Commission should lay down guidelines within three months for pricing non-firm power.[33]

Besides these general legislative initiatives definite measures are being taken to promote specific renewable energy sources. Special emphasis has been attached to promote renewable energy in urban as well as rural areas. For the urban areas the programme on Biomass Energy and Co-Generation (non-bagasse) in industry is being implemented. The major objectives of the scheme are :

- To encourage the deployment of biomass energy systems in industry for meeting thermal and electrical energy requirements.
- To promote decentralized / distributed power generation through supply of surplus power to the grid.
- To conserve the use of fossil fuels for captive requirements in industry.
- To bring about reduction in greenhouse gas emissions in industry.

---

[32] *Id* Clause 6.4 (1).

[33] *Id* Clause 6.4(2).

- To create awareness about the potential and benefits of alternative modes of energy generation in industry.[34]

The scope of the programme extends to Biomass gasifiers for thermal and electrical applications in industry, Biomass Co-generation Projects, Industrial Co-generation Projects based on Conventional Fuels and their Rejects. The programme also envisages Central Financial Assistance in the form of capital subsidy. The projects will be implemented by private and public sector industry, including through Energy Service Companies (ESCOs).[35]

Accelerated Programme for the recovery of energy/power generation from industrial and commercial wastes and effluents has been started in 2006 with the objective of assessing and upgrading various conversion technologies to mitigate the energy problems of cities and to provide alternative energy solutions for industrial and commercial establishments. The objective of the scheme also includes to accelerate the installation of energy recovery projects from industrial wastes with a view to harness the available potential by 2017.The scheme envisages Central Financial Assistance in the form of capital subsidy to promoters and grants in aid for promotional activities R&D, Resource assessment, technology up gradation etc. The scope of the scheme includes Power generation from biogas and Power generation from solid industrial waste.[36]

Various fiscal and financial benefits have been envisaged from time to time some of them consists of.

---

34 Programme on Biomass Energy and Co-generation (non-bagasse) in Industry, Ministry Of New and Renewable Energy, Government of India, available at http://mnes.nic.in/biomass-energyco.htm#becgind01(Last accessed on 12-02-2009) .

35 *Id*

36 Accelerated Programme for the Recovery of Energy Generation from Industrial and Commercial Wastes and Effluents,2006, Ministry of New and Renewable Energy, Government of India.

- 2/3rd of the project cost subject to a maximum of Rs. 2.00 crore per 100 KW for procurement of modules, structures, power conditioning units, cabling etc. to the implementing agency. The balance cost on land, extension of grid lines, transformers, civil works, foundation and erection and commissioning, etc. is met by the implementing agency.
- Up to Rs.1.0 lakh for the preparation of Detailed Project Report (DPR) for the grid interactive SPV power projects. [37]
- 2.5 per cent of its share of project cost, subject to a maximum of Rs.5 lakhs for performance evaluation, monitoring, report writing, etc. to the State Nodal Agency.
- Interest subsidy of up to 4 per cent to Financial Institutions including IREDA, Nationalized Banks etc. for captive power projects of maximum capacity 200 KW by industry.[38]

Promoting Renewable energy in rural areas assumes special significance in India. Many villages are not connected by Grid. In India around 95,000 villages have yet to be connected to electricity. Out of these, 18,000 villages may never be connected to the national grid due to their remoteness.[39] In fact India's Energy programme was started with the objective of overcoming the rural energy crisis. Various initiatives are undertaken to provide energy to rural areas. The Remote Village Electrification Programme of the Renewable Energy Ministry was initiated for electrification

---

[37] http://www.geni.org/globalenergy/library/energytrends/currentusage/renewable/Renewable-Energy-Potential-for-India.pdf (Last accessed on 3-03-2009).

[38] *Id*

[39] Renewable Energy Sources for Remote areas, available at http://data.undp.org.in/cc/ProjectBriefs/Renewable%20Energy%20Sources%20for%20Remote%20Areas.pdf (Last accessed on 3-01-2009).

through renewable energy sources.[40] The Ministry has been implementing a Remote Village Electrification (RVE) Programme since 2001-02. Renewable energy technologies such as solar photovoltaics, small hydro, biomass etc. are being deployed for electrification of remote villages and hamlets. The objective of the RVE Programme is to provide access to electricity through renewables to households in remote villages and hamlets, which are not likely to get covered by grid extension. A District Advisory Committee (DAC) on Renewable Energy headed by the District Collector is being constituted in every district to oversee the implementation of renewable energy programmes in the district.

Further under **Renewable Energy Plan, 2012,** government has proposed draft renewable energy policy and programme interventions required achieving goals of meeting the minimum rural energy needs; Plan proposes

- 10% share for renewables to set up about 12,000 MW generation capacity;
- deployment of solar water heating systems in 1million homes;
- electrification by renewables of at least one quarter of 18,000 un-electrified villages;
- deployment of 5 million solar lanterns and 2 million solar home lighting systems;
- coverage of 30 million households through improved Chullhas (wood stoves);
- setting up of further 3 million family size biogas plants.[41]

---

40 Remote Village Electrification Programme, Ministry of New and renewable energy, Government of India.

41 RENEWABLE ENERGY PLAN 2012 , NATIONAL RENEWABLE ENERGY POLICY , Ministry of New and Renewable Energy, Government of India.

Another significant aspect of the plan relates to the emphasis ascribed to female participation for their employment and empowerment; availability of minimum cooking energy to all house holds; to provide cost effective energy for water pumping, irrigation, drinking and for rural electrification and all round rural development through the Integrated Rural Energy Programme (IREP).[42]

Creating public awareness has always been on the agenda of various government policies. In order to give a fillip to this, The Renewable Energy Park Scheme was started in 1994-95 under the Special Area Demonstration Programme.[43] The main objective of the Renewable Energy Park Scheme is to create awareness, publicity and provide an opportunity to the students / teachers and rural and urban masses about the use and benefits of the renewable energy by demonstrating new and renewable energy systems and devices by demonstrating working systems, cut models, L E D models, blow ups etc.[44] Two types of Renewable Energy Parks are being supported under this scheme namely District Level Renewable Energy Parks(DLEP) and State Level Renewable Energy Parks (SLEP).[45] The Renewable Energy Parks have been found quite successful in demonstrating NRSE systems and technology creating awareness among target groups and more or less self sustained in recurring cost.[46]

---

42 Integrated Rural Energy Programme (IREP) was conceptualised during the Sixth Five Year Plan and launched as Centrally Sponsored Scheme in the Seventh Plan. IREP was transferred in 1994-95 from the Planning Commission to MNRE. Major object is to provide for minimum domestic energy needs for cooking, heating, and lighting purposes to rural people in selected village clusters in each District, with focus on renewable energy.

43 Ministry of New and Renewable Energy Information and Public awareness Group Special Area Demonstration Programme available at http://mnes.nic.in/energy-park-scheme.htm(Last accessed on 10-02-2009).

44 *Id*

45 *Id*

46 *Id*

The Indian Renewable Energy Development Agency has been set up under Ministry for Non-Conventional Energy Sources and is a specialized financing agency to promote and finance renewable energy projects.[47] Generally in India renewable energy policies were more or less financial. Fiscal incentives are normally aimed to encourage utilities to buy renewable power, promoter companies to set up renewable projects, equipment companies to manufacture renewable equipment or private and government entities to undertake R&D relating to renewable. In India, policy initiatives encourage domestic private as well as FDI investments with a provision of fiscal and financial incentives such as tax holidays, accelerated depreciation, soft loans, non imposition of excise duties and duty rebates.

Apart from these fiscal steps recently certain emergency measures are taken. Coal-based power plants account for about two-thirds of the total electric generation and government is trying to link tariffs to efficient management. Plants are being encouraged to adopt more efficient and clean coal technologies, an energy labeling programmed for appliances was launched in 2006, and the labels provide information about the energy consumption of an appliance like Air Conditioners.[48] An Energy Conservation Building Code (ECBC) was launched in May, 2007, which addresses the design of new, large commercial buildings to optimize the

---

[47] IREDA is a Public Limited Government Company established in 1987, under the administrative control of Ministry of New and Renewable Energy (MNRE) to promote, develop and extend financial assistance for renewable energy and energy efficiency/conservation projects with the motto http: // www.geni.org/globalenergy/library/energytrends/currentusage/renewable/Renewable-Energy-Potential-for-India.pdf.

[48] "National Energy Labeling, Bureau of Energy Efficiency, Ministry of Power has developed a scheme for energy efficiency labeling of equipment, Participation in the scheme is voluntary and currently applicable for the following equipment: 1. Frost-Free (No-Frost) Refrigerators and 2. Tubular Fluorescent Lamps.

building's energy demand.[49] Under the Energy Conservation Act 2001, the conduct of energy audits was made mandatory in large energy-consuming units in industrial sectors.[50] Several Commissions have already operationalized this mandate, and also notified preferential prices for electricity from renewables. This has contributed to an acceleration in renewable-electricity capacity addition, and over the past three years, about 2,000 MW of renewable-electricity capacity has been added in India every year, bringing the total installed renewable capacity to over 11,000 MW.

International mechanisms to promote energy efficiency and production of clean energy also needs a mention here. The Kyoto protocol has created a favorable climate for attracting international support for renewable energy development. The Clean Development Mechanism (CDM) was one of the avenues through which financial resources can be raised and sustainable development can be promoted.[51] As on 17 March 2009, 398 out of total 1455 projects registered by the CDM Executive Board are from India, which is next only to China with 453 projects. The National CDM Authority (NCDMA) in India has accorded Host Country Approval to 1226 projects facilitating an investment of more than Rs.151,397 crores. These projects are in the sectors of energy efficiency, fuel switching, industrial processes, municipal solid waste and

---

[49] Energy Conservation Building Code (ECBC) The Energy Conservation Building Code (ECBC) sets minimum energy performance standards for commercial buildings. The code has been prepared pursuant to section 14 (p) of the Energy Conservation Act, 2001, which empowers the Central Government to prescribe ECBC for commercial buildings and building complexes for efficient use of energy and its conservation. The state governments have the flexibility to amend ECBC to suit local or regional needs.

[50] Energy Conversation Act, 2001.

[51] The Clean Development Mechanism (CDM) is an arrangement under the Kyoto Protocol allowing industrialised countries with a greenhouse gas reduction commitment (called Annex 1 countries) to invest in projects that reduce emissions in developing countries as an alternative to more expensive emission reductions in their own countries.

renewable energy. If all these projects get registered by the CDM Executive Board, they have the potential to generate 573 million Certified Emission Reductions (CERs) by the year 2012.[52]

In recent times Major thrust is being given towards - research, development, demonstration, commercialization and deployment of new and renewable energy systems/devices for transportation, portable and stationary applications for rural, urban, industrial and commercial areas. This is a shift from the subsidy driven and financial incentive led programme. Bio-fuel is being developed for motive power for stationary/portable applications. The government is also keen to launch a National Mission on Biodiesel, using about 11 million hectares of land to produce biodiesel by 2011-2012.[53] For example, India's National Auto-fuel Policy mandates cleaner fuels for vehicles.[54] National Hydrogen Energy Board has been set up under Ministry of New and Renewable Energy. The Road Map of the energy board has called for the adoption of a total systems approach for developing hydrogen energy technologies though public private partnership.[55]

One of the significant policy step in the direction of renewable energy can be find in the National Action plan on climate change released in 2008. The plan envisages the establishment of National solar Mission and National Mission for Enhanced Energy Efficiency both aiming at augmenting renewable: The Act plan emphasis a system for companies to trade energy-savings certificates; Energy incentives, including reduced taxes on energy-efficient appliances; and Financing

---

[52] CDM India, available at http://cdmindia.nic.in/(Last accessed on 23-02-2009).

[53] National Mission on Bio Mass Energy 2006.

[54] National Auto fuel Policy 2003, The policy lays down a roadmap for achieving various vehicle emission norms by 2010.

[55] National Hydrogen Energy Board, Ministry of New and Renewable Energy Government of India 2006.

for public-private partnerships to reduce energy consumption through demand-side management programmes in the municipal, buildings and agricultural sectors. An exclusive and comprehensive policy on renewable has been proposed, aiming to raise renewable energy capacity to 100,000 MW by 2050.[56]

Further the integrated energy policy of India, which has been in the making since 2005, finally received Cabinet approval in the last week of December 2008. The integrated energy policy of India 2009 marks a new beginning in India's renewable energy policy.

The Integrated Energy Policy of India envisages electricity generation installed capacity of 800 000 MW by 2030 and a substantial contribution would be from renewable energy. Salient features of the new policy consists of encouraging power generation from renewable sources. The policy lays down time frame to increase the share of renewable energy to 10% by 2010 and 20 % by 2020. In order to achieve this policy envisages a supportive fiscal regime, single window clearance, leveraging additional budgetary resources from other departments, and preferential prices for renewable electricity.

Policy proposes setting up of separate time-bound Technology Missions Solar Energy, Bio fuels and Hydrogen to achieve the objectives of energy independence.[57] The solar mission will initiate a time-bound programme of demonstration of solar rooftop lighting systems in 10,000 government buildings by 2010 and will promote solar water heating mandatory throughout the urban areas of the country by 2012, in a phased manner. The Technology mission for Bio fuels aims at time-bound conversion of the present about 18,000 MW diesel-based captive generating units to bio-diesel-based units. It also provides for small biomass based energy

---

56 National Action Plan On Climate Change, India 2008.

57 Integrated Renewable Energy Policy, India 2008.

systems for rural areas. Special emphasis is also given to develop road map for developing a hydrogen and fuel cell economy.

The policy envisages a national level apex body called the Renewable Energy Council with the Central Minister for NRE as its Chairman with substantial representation from Industry, Academia, Non-Government Institutions, Researchers and User-Groups. As a complementary measure plan proposes participation and involvement of National Research Laboratories, IITs, Universities, Industry, Specialist, Non-Government Institutions and User Groups actively in RE decision making and research efforts.

At present, the Union government can issue only guidelines and recommendations to the state governments. The Act would enable the Centre issue mandatory orders, the Union minister of state for non-conventional energy sources. Different states are having diverse policies which undergo frequent changes confusing the investors and users of renewable energy. The new Act will be an answer to this problem and enable the country to achieve its goal of 10,000MW renewable energy capacity by 2012 if implemented properly.A key proposal of the policy that needs to be supported is that of tax neutrality across energy sources. This combined with full cost pricing should remove the grouse of the renewable energy industries regarding the absence of a level playing field to a large extent and give us a true estimate of cost difference between renewable energy and conventional energy forms.

Renewable Energy Policies in India: Too many promises to deliver

During the last two decades India's renewable energy programmes have mature in volume, and reach. Initially, the key of the national effort was aimed at capacity building and promoting research and development, mostly in national laboratories and educational institutions. However, a major policy shift was noticed and a huge expansion was witnessed in activities from the 1980s onwards, The focus shifted to large-scale demonstration and subsidy-driven extension activities mainly in providing energy services to rural areas through biogas, improved cooking stoves and solar energy. These programmes helped created awareness, generated field experience, and helped set up a vast network of institutions and non-governmental organizations reaching right down to self-employed workers at the grassroots level. The stress of policy is currently on commercialization with private sector participation in power generation from wind, small hydro and biomass combustion/gasification, as well as in industrial applications of solar and other forms of renewable energy.[58]

Despite India being the only country with a dedicated Ministry for renewable energy, and an extensive set of policies, the renewable energy sector is witnessing slow growth. Key reasons include multiplicity of agencies, skewed incentive structure, and poor implementation capability.[59] There is multiplicity of government-led agencies in the renewable energy sector, leading to lack of accountability towards the growth of the sector. For instance, while MNRE plans, implements, and manages incentives and support; Department of Science and Technology is responsible for research and development. Further, Ministry of Power (MoP) is responsible for designing policies for grid-connected power supply from renewable energy projects. Also, all states have different policies regarding renewable energy projects. This is a big deterrent as promoters have to separately negotiate with each state department. Although considerable experience and

---

58 Preety Bhandari, India's pragmatic approach to climate change,31,August 2006.

59 V Subramanian, Renewable energy in India: status and future prospects,available at < http://www.climateactionprogramme.org/features/ article/renewable_energy_in_india_status_and_future_prospects/ (Last accessed on 4–3-2009).

capabilities exist on renewable electricity technologies including the development of indigenous biomass gasifier technology and manufacturing base for wind power and solar photovoltaic, a number of barriers still remain to be overcome. The push policies adopted since the nineties have been successful in creating a fairly large and diversified manufacturing base and an infrastructure (technology-support groups and facilities, as well as the nodal agencies) to support RET design, development, testing, and deployment. But commercialization of the technologies have been limited due to low reliability of the devices, lack of remunerative tariffs for RET-generated electricity, and a lack of consumer-desired features (in terms of the services and the financial commitments) in the design and sales-package. Further Distortions in the energy and electricity prices and non internalisation of the socio-environmental externalities have impeded the progress of RETs by adversely affecting their competitiveness compared to conventional energy sources. With regard to rural renewable energy programme the obstacles includes Low per capita income, Non-availability of upfront capital, Lack of micro-finance mechanisms, Remote locations, Infrastructure for initial installations, Availability of skilled persons Operation and Maintenance and Repair & replacement mechanism.

Another major concern is the availability of renewable energy at an affordable price. At present generation of renewable energy are not very much financially viable compared to fossil fuel resources. Indeed the entire scope and future of renewable energy will revolve around this issue. E.g. Cost of producing power from solar panels is higher than the conventional thermal power plants: Conventionally generated electricity ranges between 2 and 4 Rupees per kilowatt hour (the amount of money to get a kilowatt of power for an hour). Solar PV and thermal costs around 10 to 15 Rupees a kilowatt hour. Even with the incentives, it will be tough for the solar companies to run the solar power plants on sustainable basis.

Also, most of the equipments need to be imported from outside. Indian companies do not have the skill set for design, built and operate the solar power plants. These factors inhibit the development of renewable energy in India and need immediate attention from government.

Further any policy on renewable energy should aim at achieving sustainable development. The main principles of a sustainable energy policy must fulfill the following criteria:

- A mix of centralized and decentralized energy systems to meet the needs of urban and rural India.
- Emphasis on energy efficiency throughout its life cycle. Energy audits have shown a 25 per cent reduction reduction in energy demand can be met by good house keeping.
- Correct pricing of energy and a level playing field for all forms of energy. The distortions due to subsidies and taxation skew the pricing.
- Renewable energy use wherever possible to meet specific and niche needs of urban, rural, remote, and hilly areas.[60]

Further equal emphasis should be given to Private sector participation as they can contribute not only capital but can promote technology too. The Green Business Center (GBC) at Hyderabad, India is a good example of a partnership between the state government of Andhra Pradesh, the Confederation of Indian Industry and USAID devoted to showcasing of energy efficient technologies, dissemination of information on "green" technologies and providing green certification.[61] Renewable energy-based enterprises, unlike the mainstream

---

60 Dr Arcot Ramachandran, "Renewable Energy Micro-Grids for Rural India", *Renewable Energy India*, Vol 3, p 29,(2005).

61 *Supra* note 23

energy sources, are relatively decentralized both at the level of sources as well as end-uses. Therefore, entrepreneurship development in this sector requires local decision making and promotion of local stakeholders.[62]

The Indian government can play an important role in encouraging the growth of alternative energy sources, not only by giving cash subsidies but through other kinds of incentives. For instance, wind energy received a boost in Tamil Nadu when the Ministry of Textiles agreed to include wind farms in its Technology Upgrading Fund Scheme, which gives a subsidy of five per cent in the rate of interest on capital borrowed for upgrading textile mills. As a result, several textile mills in the state set up wind energy farms. Even without such incentives, some industries have already recognized the potential saving from tapping the wind. Bajaj Auto, a company producing two-wheelers in the city of Pune, south of Mumbai, has set up a wind farm that writes off its entire electricity bill; several other industries are following its example.[63] This also depicts the need to promote public and private sector cooperation.

CONCLUSION

Climate change and its consequences are ubiquitous. In order to mitigate the effects of climate there should be overall reduction on carbon centric energy pattern. A uniform, national policy note on renewable energy will go along way in achieving this target. Such a sustainable comprehensive energy policy has a two prolonged strategy. It will not only meet the future energy demand for rapid economic growth but also protect the environment and conserve scarce resources. In addition to that renewable energy policies will be

---

[62] P.M. Mathew, "Need For a New Paradigm Shift: Mainstream Development Strategies fail to address issues of Entrepreneurship", *Renewable Energy India*, Vol 3, p 17,(2005)

[63] Kalpana Sharma, *"Blowing in the Wind"* Data Paper 6- 53rd Commonwealth Parliamentary Conference, India, 2007,available. athttp://www.cpahq.org/uploadedFiles/Programmes_and_Activities/Professional_Development/DP%206.DOC(Last accessed on 5-03-2009).

a positive step toward mitigating climate change phenomenon. The potential benefits of participating in efforts to address Climate Change through renewable energy are available now and in the immediate future. India could take advantage of these efforts to address Climate Change and thereby increase its foreign direct investment, technology transfer, and job creation while also taking a responsible step towards reducing environmental impacts. In this context, the Indian renewable energy programmed needs to a goal-oriented effort to meet the country's energy requirement in an environmentally sound way. The challenge is formidable indeed for renewable technologies which, compared to fossil fuel technologies, remain primitive, inefficient and uneconomical at present.

4

Solar Power Policy in India

A Scientific, Environmental and Legal Analysis

"In this (India's) strategy, the Sun occupies centre stage, as it should, being literally the original source of all energy. We will pool all our scientific, technical and managerial talents, with financial resources, to develop solar energy as a source of abundant energy to power our economy and to transform the lives of our people. Our success in this endeavor will change the face of India".

—India's Prime Minister, Dr. Manmohan Singh, releasing the National Action Plan on Climate Change in 2008.

INTRODUCTION

India is facing a huge power demand-supply gap. The situation has assumed greater criticality in lieu of rising fuel price and the larger requirements created by economic development. It is imperative that to sustain our economic growth and to promote the welfare of people the power demands needs to be fulfilled. Power Sector Reforms are will play an important role in rapid development of India. This single issue alone may play a vital role in deciding how fast India can take to the category of developed country from the current developing country status. The chapter looks at solar energy as an option to overcome the power crisis in India. The chapter discusses in details the potential of solar power in India and analyses associated rules and policies initiated by government of India.

UNSUSTAINABLE ENERGY CONSUMPTION PATTERN

In India and world, fossil fuel remains the main stay of energy generation. Coal remains the main source of electricity generation in India due to vast native resources and constitutes about three-fifths of the country's power generation capacity. (Table 4·0) Large hydropower projects at present have about one-quarter share, and has consistently showing declining trend over the past decade. Its growth has remained restricted primarily due to a number of socio-environmental barriers. Gas-based power has grown very rapidly in the last decade to almost one-twelfth of total generation capacity due to lower capital requirements, shorter construction periods, and higher efficiencies. However, Gas based power has been unable to meet the exponential growth in demands of energy because of limited supply of gas. Nuclear capacity is about four per cent of the total energy

Table 4·0: Installed Power Generation Capacity in India, 1997-2008 (in GW)

	1997	*2002*	*2003*	*2004*	*2005*	*2008*
Hydroelectric	21.65	26.26	26.76	29.5	30.94	36.7
Nuclear	2.22	2.72	2.72	2.72	2.77	4.12
Solar/Wind/Bio-mass/Renewables	1.27	1.51	1.74	1.87	3.81	13.45
Conventional Thermal	59.64	74.55	76.65	77.97	80.9	93.4
Total Capacity	85.79	105.05	107.88	112.06	118.42	147.6

Sources: DOE/EIA, Carbon Sequestration Leadership Forum, Ministry of1 Power, Ministry of New and Renewable Energy (MNRE)

generation.[1] There are environmental and safety concerns which inhibit substantial growth in nuclear energy capacity. The renewable energy including Solar, Wind and Biomass contributes very less in the total share at present.

The situation is similar worldwide. In 2007, according to the International Energy Agency, 80.3% of the world's energy came from fossil fuel: Oil (34.3%), Coal (25.1%) and Gas (20.9%).[2] Overall, 90.9% of the world's energy came from combustion, because alongside these fossil fuels in 4th place are "combustible renewables," mostly wood (10.6%). If we include nuclear power (6.5%) and hydro-electric power (2.2%) in above, this will accounts for 99.5% of the world's energy. Wind and solar, currently only provide .064% and .039% of the world's power requirements. For solar energy to achieve its potential and replace all other sources of energy in the world, this .039% would have to increase 2,500 times.

Table 4·1: World Power Profile

Category	*Percentage share (2006)*
Oil	34.3%
Coal	25.1
Gas	20.9
Combustible	10.6
Nuclear	6.5
Hydro	2.2
Wind	.064
Solar	.039

Source: Author, Data IEA

Current energy pattern poses three major problems; firstly the present energy consumption based on fossil fuel is not

[1] Shukla, P. R., Pierre, Audinet and Grare, Frederic, (Eds.) India's Energy: Essays on Sustainable Development. Manohar Publication. New Delhi, (2000).

going to undergo a substantial change in the near future. It is estimated that the world energy demand expands by 45% between now and 2030 -an average rate of increase of 1.6% per year -with coal accounting for more than a third of the overall rise.[3] Further since the industrialized nations only comprise approximately 20% of the world's population yet consume over 50% of the world's energy production, it is unlikely that global energy production will not have to increase. Secondly continued reliance on fossil fuel as energy resource is likely to lead to increasing emissions of gases, compounding the pollution problems at the local level and increasing Green House Gas (GHG) emissions resulting in climate change. For instance India power sector alone has a 40 per cent contribution in the total carbon emissions.[4] A long term projection of the business-as-usual scenario over a forty year period (1995-2035) indicates that energy consumption shall treble; electricity generation shall rise by 5.4 times; coal shall continue to be the main source of fuel and carbon emissions shall go up by 3.6 times.[5] Last and not the least these resources are clearly depleting and will not be lasting more than a couple of decades to cater to the increased energy demand for decades to come. Thus Current energy trends are unsustainable, socially, environmentally, economically. Oil is likely to remain the leading energy source. To avoid "abrupt and irreversible" climate change and other related environmental crisis we need a major decarbonisation of the world's energy system and that shift can be provided by renewable energy sources. Using sunlight to create electrical and thermal energy remains the most promising source of clean renewable energy, and projections as to how quickly

[3] World Energy Outlook,2008,IEA.

[4] *Id*

[5] Shukla, P. R and Pandey, R. "Energy and Emissions Projections for Indian Industries using AIM\ENDUSE Model", in Energy and Emissions Studies using AIM Model - Eds. T. Morita, Y. Matsuoka and M. Kainuma, Springer-Verlag Publications,(1999).

solar power takes off depends on various technical and financial factors. The present economic recession do not excuse back-tracking or delays in taking action to address energy challenges.

SOLAR ENERGY: APPLICATION AND BENEFITS

Solar power is the technology of obtaining usable energy from the light of the sun. Solar power is an extremely efficient and clean way to generate electricity. Energy from the sun has many features, which make it an attractive and sustainable option: global distribution, pollution-free nature, and the virtually inexhaustible supply. Solar energy technologies offer certain socio economic and environmental advantages. Environmentally there is no air emissions associated with the operation of solar modules or direct application technologies. Residential-scale passive construction, photovoltaic, solar water heating, and other direct applications reduce power generation from traditional sources and the associated environmental impacts.[6] Further regarding the environment, the use of SETs has extra positive implications such as:

- Reduced emissions of the greenhouse gases (mainly CO_2, NOx) and avoidance of toxic gas emissions like SO_2, particulates;
- Encourages reclamation of degraded land;
- Reduction of the required transmission lines of the electricity grids; and
- Improvement of the quality of aquatic resources.

Various socio economic benefits are claimed

- Decrease the energy crisis;
- Increase of the regional/national energy independency;

---

[6] Solar Energy Technology, Renewable Northwest project available at http://www.rnp.org/renewtech/tech_solar.html(Last accessed on 2-03-2009).

- Momentous generation of job opportunities;
- Diversification and environmental security of energy supply;
- Assist the deregulation of energy markets; and
- Acceleration of the rural and inaccessible are electrification in developing countries.

TWO ROUTES TO HARNESS SOLAR ENERGY

Solar energy can be harnessed through two routes, namely solar photovoltaic and solar thermal, by direct conversion to electricity and heat energy respectively.

SOLAR PV POWER GENERATION TECHNOLOGIES

PV generation technology is commercially proven and large multi-megawatt generation plants have been operating since the 1990s. Costs associated with the technology are high, but the technology is well-known and reliable. The largest plants are based on fixed solar panels inclined at latitude angle. Across the world, this has proved to be the most economic way of building PV power

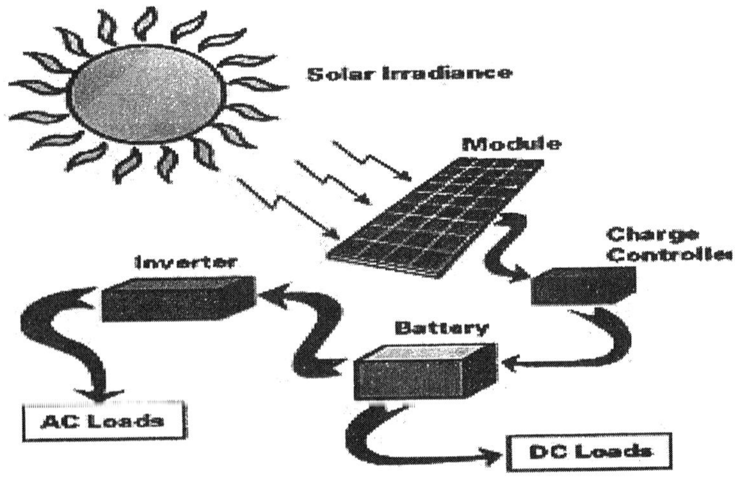

Figure 4·1: Schematic of Solar PV System

stations. More recent developments use PV collectors that track the Sun to allow collection of a greater amount of energy and concentrating photovoltaic (CPV systems that focus the collected solar energy into a smaller area).

SOLAR THERMAL POWER GENERATION TECHNOLOGIES

Solar Thermal Power systems, also known as Concentrating Solar Power (CSP) systems, use concentrated solar radiation as a high temperature energy source to produce electricity using the thermal route. Since the average operating temperature of stationary non-concentrating collectors is low (max up to 120°C) as compared to the desirable input temperatures of heat engines (above 300°C), the concentrating collectors are used for such applications. These technologies are appropriate for applications where direct solar radiation is high. The mechanism of conversion of solar to electricity is fundamentally similar to the traditional thermal power plants except use of solar energy as the source of heat.

Solar collectors are used to produce heat from solar radiation. However, the three most promising CSP technologies are parabolic trough, central receiver or solar tower, and parabolic dish.[7]

(*a*) Parabolic trough system: Temperature of the receiver can reach 400° C and produce steam for generating electricity. This is conducted along a heat exchanger in which steam is produced, which then generates power in the turbines.[8]

(*b*) Power tower system: The reflected rays of the sun are always aimed at the receiver, where temperatures well above 1000° C can be reached.

[7] Kourosh Sedghisigarchi, Asad Davari, Concentrated Solar Power: Technology and Economic Study, Emerging Energy Research,(2006).

[8] Concentrating Solar Power from Research to Implementation", Directorate - General for Energy and Transport, European Commission, (2007).

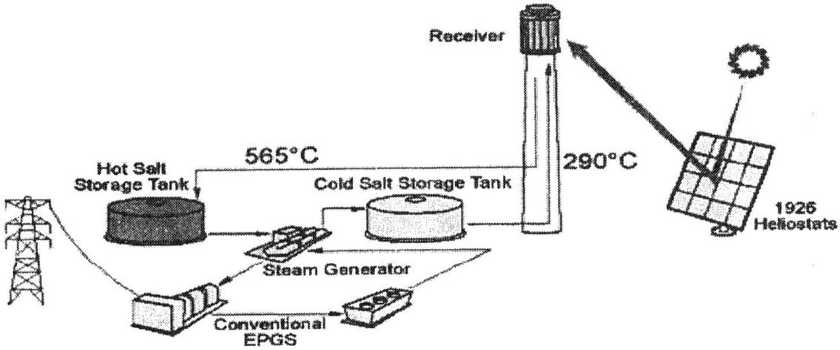

Figure 4·2: Schematic of Power Tower System

(*c*) Parabolic dish systems: Parabolic dish systems can reach 1000° C at the receiver, and achieve the highest efficiencies for converting solar energy to electricity.

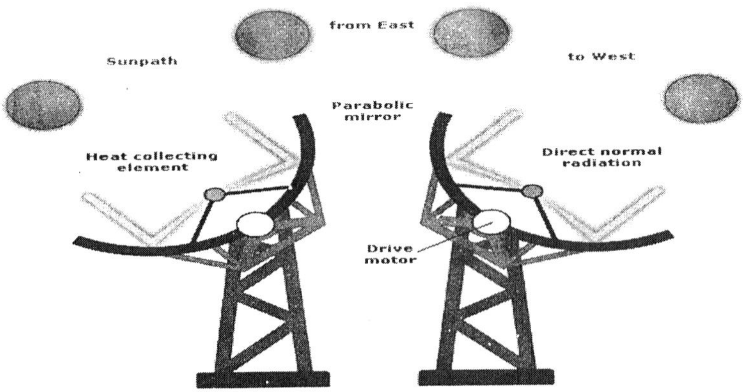

Figure 4·3: Parabolic Dish Systems

Variations of the annual solar irradiation depending on the site influence these technologies annual efficiency, specific output and electricity generation cost. Solar energy, experienced by us as heat and light, can be used through these two routes: the thermal route uses the heat for water heating, cooking, drying, water purification, power generation, and other applications; the photovoltaic route converts the light in solar energy into electricity, which can then be used for a number of purposes such as lighting, pumping, communications, industrial and power supply in unelectrified areas.[9] These technologies generally find applications in various products. Some of the prominent solar energy products and uses include,

- Solar collectors which gather the sun's energy, transform its radiation into heat, then transfer that heat to water, solar fluid, or air.
- Solar water heaters are a cost-effective way to generate hot water. They can be used in any climate.
- Solar cells are materials that turn sunlight into electricity.
- Solar Panels are a form of active solar power, solar panels harvest sunlight and actively convert it to electricity.
- Solar cooker is a device, which cooks food with the help of solar energy and can save the conventional fuels to a significant amount.[10]

SOLAR POWER STATUS AND POTENTIAL IN INDIA AND WORLD

The buzzword in any environment discourse today is climate change. A responsible climate change policy should take into account the principles of energy conservation and efficiency. Renewable energy offers both these benefits. At present renewable energy contributes to 17 per cent of the world's primary energy consumption. But these include Biomass, Hydropower and new Renewable energy sources in the form of Wind, Solar, etc.

Solar energy contributes a mere 0.39% to the total renewable energy.

---

[9] Reuel Shinnar, Francesco Citro, "Solar thermal energy: The forgotten energy source", Technology in Society 29,261-270,(2007).

[10] Solar Energy Info-kit, TERI, Envis Center on Renewable Energy and Environment p 33, 2008.

Table 4·2: Renewable Energy Consumption by Energy use 2002-2006 (Quadrillion Btu)

Sector and Source	*2002*	*2003*	*2004*	*2005*	*2006*
Total	5.893	6.150	6.261	6.444	6.922
Solar/PV	0.065	0.063	0.065	0.067	0.072
Residential Solar/PVh	0.059	0.058	0.059	0.061	0.067
Electric Power Solar/PV	0.006	0.005	0.006	0.006	0.005

Source: IEA 2006

The growth of solar energy sector was happening from 1970. according to the IEA's factsheet, "Renewables in Global Energy Supply," the solar energy sector has grown by 32 per annum since 1971 and Solar Energy demand has grown at about 25 per cent per annum over the past from 1985 to 2000.[11] In 1985, worldwide annual solar installation demand was only 21 MW. Regarding the solar market the fastest growing energy technology in the world has been grid-connected solar PV, with total existing capacity increasing from 0.16 GW at the start of 2000 to 1.8 GW by the end of 2004, for a 60 per cent average annual growth rate during the five-year period. Renewable Global Status Update Report 2006 confirmed the continued growth of Grid-connected solar PV, with a 55 per cent increase in cumulative installed capacity to 3.1 GW, up from 1.8 GW in 2004.[12] During the same period, other renewable energy technologies grew rapidly (annual average) as well: wind power 28 per cent, bio-diesel 25 per cent, solar hot water/heating 17 per cent, off-grid solar PV 17 per cent.[13]

Grid-connected solar photovoltaic (PV) continued to grow fast with a 70- per cent increase in existing capacity to 13 GW in 2008. This represents a six fold increase in global capacity

---

11 Renewables in Global Energy Supply," Fact sheet IEA,2006.

12 Renewables, Global Status Report, 2006 Renewable Energy Policy Network for the 21 st Century.

13 *Id*

since 2004 (Fig. 4·4). Annual installations of grid-tied solar PV reached an estimated 5.4 GW in 2008.[14]

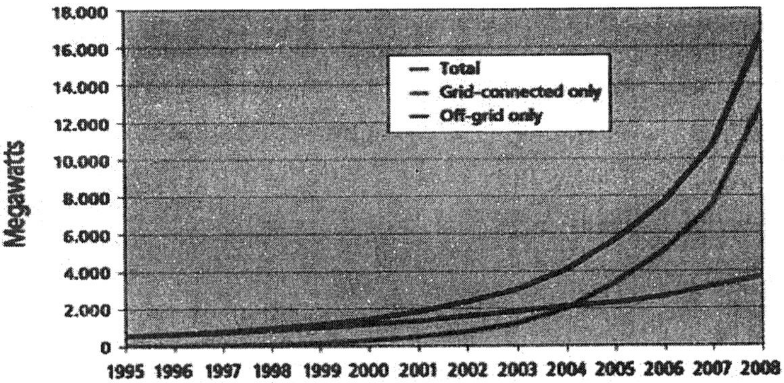

Figure 4·4: Solar PV, Existing World Capacity, 1995-2008

Source: Global Status Report Renewable 2009

In the assessment of PV demand in 2008, 81 countries contributed to the 5.95 GW world market total. Europe accounted for 82% of world demand in 2008. Spain's 285% growth pushed Germany into second place in the market ranking, while the US advanced to number three. Rapid growth in Korea allowed it to become the fourth largest market, closely followed by Italy and Japan (Fig. 4·5) .

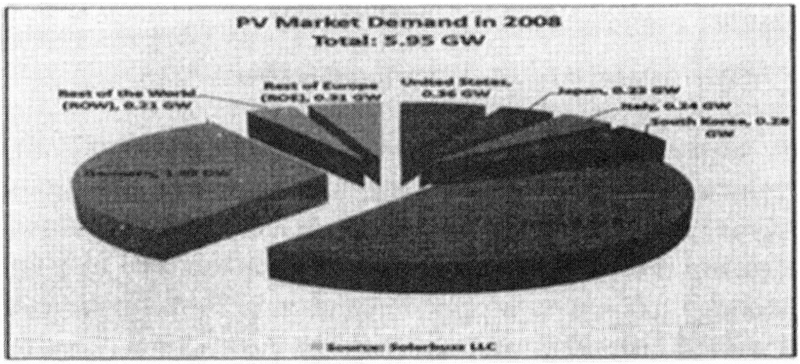

Figure 4·5: PV Market Demand in 2008

---

[14] *Id* Renewables Global Status Report 2009.

On the supply side, world solar cell production reached a consolidated figure of 6.85 GW in 2008, up from 3.44 GW a year earlier. Overall capacity utilization rose to 67% in 2008 from 64% a year earlier. Japan was the leader in cell production (830 MW), followed by Europe (470 MW), China (200 MW), and the US (150 MW).But in 2009 China usurped Japan to become the new world leader in PV cell production.[15] Solar hot water in Germany set record growth in 2008, with over 200,000 systems installed.[16] But China continues to occupy the maximum share (Fig. 4·6).

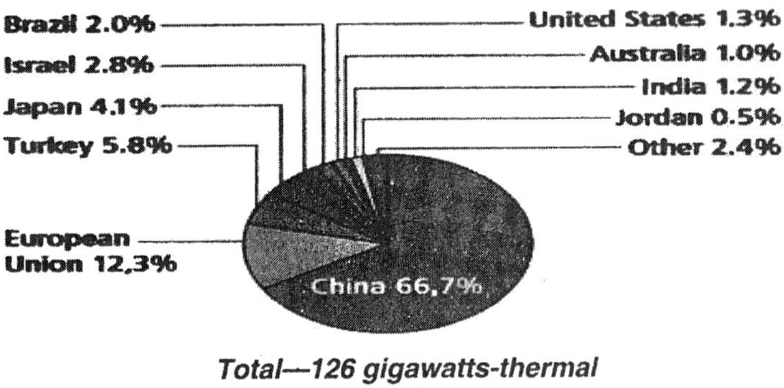

Figure 4·6: Share of Solar Hot Water/Heating Capacity Existing, Top 10 Countries, 2007

Source: Global Status Report Renewable 2009

Regarding India the future economic development process for India is likely to result in rapid and accelerated growth in energy demand. Renewable can replace fossil fuels for reducing the greenhouse gases emissions and also help in meeting the power supply-demand gap. India possesses renewable energy sources in abundance the potential capacity

15 *Id*

16 *Id*

is estimated to be 126,000MW.[17] Despite these potential only 8 per cent of India's energy comes from renewable sources.[18]

India due to its geo-physical location receives solar energy equivalent to nearly 5,000 trillion kWh/year, which is far more than the total energy consumption of the country today. The equivalent energy potential is about 6,000 million GWh of energy per year. The estimated potential of power generation through solar photovoltaic system is about 20 MW / Sq.km in India.[19] Fig. 4·7 shows map of India with solar radiation levels in different parts of the country.

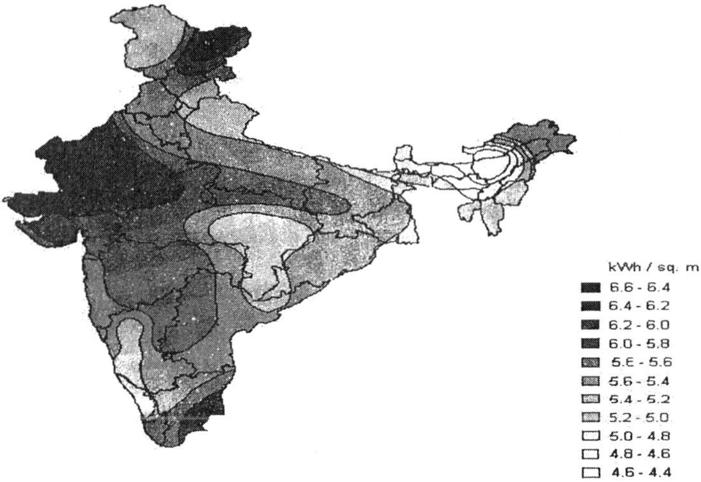

Figure 4·7: Map of India with Solar Radiation Levels

Source: TERI

---

[17] Raghuvanshi, S. P., Chandra A. and Raghav A. K., "Carbon Dioxide Emissions from Coal based Power Generation in India" - Energy Conversion and Management 47: 427-441,(2006).

[18] Shoumyo Majumdar, The Current Scenario Of Developments In Renewable Energy In India, REEP 2008.

[19] Prabhakant and G.N. Tiwari, "Energy payback time and life-cycle conversion efficiency of solar energy park in Indian conditions"- International Journal of Low-Carbon Technologies, doi:10.1093/ijlct/ctp020

It can be observed that although the highest annual global radiation is received in Rajasthan, northern Gujarat and parts of Ladakh region, the parts of Andhra Pradesh, Maharashtra, Madhya Pradesh also receive fairly large amount of radiation. India has significant advantages it can leverage in promoting solar energy as an alternative and renewable energy source. Firstly, the intensity and duration of sunlight available on its landmass is relatively greater than in many other heavily populated regions of the world.[20] Just 1 per cent of India's land area can meet its entire electricity requirements up to 2030. Secondly, the country's very size permits a rapid build-up of capacity to levels that can allow significant economies of scale. Another factor adds to the attractiveness of solar power, even at current technological levels. India's peaking power requirements during the daytime hours provide a virtually precise fit to the availability of solar power through photovoltaic.[21] So, India has huge potential to improves its position in world in the production of solar power because of it's geographical location.

But India produces a very negligible amount of solar energy - a mere 0.2 per cent compared to other energy resources. Power generation from solar thermal energy is still in the experimental stages in India. Solar energy, remains hugely underutilized despite its vast potential in a country where many areas get sunshine for most of the year. However, solar power could serve as an excellent means of bridging the huge power gap of India. Acute shortage of supply coupled with peak load energy crisis in urban areas may be tackled to a great extent by integration of Solar PV into grid system.

Up till now, India's energy base has been more on conventional energy like coal and oil. Solar energy programme begun in the mid 70's in India. While the world has progressed substantially in production of basic photovoltaic cells, India has fallen short to achieve the worldwide momentum. Our progress has been slow even in

20 Syam Saran, "Mission Possible" OUR PLANET Generating Power, Jobs and Developments,(2008).

21 *Id*

other solar technologies and solar products. Only after 1997 it started picking up. In early 2000, nine Indian companies were manufacturing solar cells.[22] The total installed manufacturing capacity was estimated to be 19 MW per year. More than 1.2 million solar lighting systems have been installed in the country. The latest, 100 million dollars investment from Tata BP Solar in India is the pointer towards the booming solar market in India. Of late, the market is growing for SPV applications based products with the active encouragement of the government.[23]

India has now attained 7th place worldwide in Solar Photovoltaic (PV) Cell production and 9th place in Solar Thermal Systems. Grid-interactive renewable power installed capacity as on 31.10.2008 aggregated 9,231 MW corresponding to around 8 per cent of the total power installed capacity which equates to over 2 per cent of total electricity. Solar photovoltaics (SPV) with an aggregate capacity of 47 MWp, has a two and a half percent contribution in the renewable based power generation capacity.[24]

What does it signifies

PV Based Systems	Total Installation
Solar Street Lighting systems	54,795
Home Lighting Systems	434,692
Solar Lanterns	697,419
Solar PV pumps	7,148
Solar PV Generation Plants	2.12 MWp

Source: MNRE Website Data, January 2009

[22] Ron Mahabir, Indias Solar Power, available at http://asiacleantech.wordpress.com/2007/10/22/indias-solar-power/(Last accessed on 10-03-2009). See also Elena,Indias Solar Dream,available at http://www.off-grid.net/2007/05/16/india%E2%80%99s-solar-dream/(Last accessed on 5-02-2009).

[23] Supra note 11

[24] Shukla P.R., Ghosh D., Chandler W., Logan J., Developing Countries and Global Climate Change: Electric Power Options in India- Prepared for the Pew Centre on Global Climate Change, Arlington, US,(1999).

SOLAR ENERGY PROGRAMME IN INDIA

Solar energy can be exploited for meeting the ever-increasing requirement of energy in our country. Its suitability for decentralized applications and its environment-friendly nature makes it an attractive option to supplement the energy supply from other sources. Development efforts to use solar energy started in the country quite early. Over the years, the country has developed an extensive research and expertise, a rising manufacturing competence, and a pervasive infrastructure for distribution and promotion of solar energy products. Since its inception, the Ministry of Non-Conventional Energy Sources (MNES) has been implementing various solar energy programme with the objective of providing electricity to the rural and remote areas.[25] Solar energy policy in India is majorly in the form of financial incentives, regulatory incentives and policy initiatives.

Ministry of Renewable energy encourages P.V technology as well as Thermal technology. A country wide Solar Photovoltaic Demonstration Programme is being implemented by MNES for more than two decades. The Programme is aimed at developing cost effective PV technology and its applications for large-scale diffusion in different sectors, especially in rural and remote areas. Major components of the solar PV programme include, R&D, Demonstration and Technology Utilization, Testing and Standardization, Industrial and Promotional activities etc.[26]

Ministry has also been implementing solar thermal power generation programme having the following components:

- Solar cooking: The programme is in the form of financial support for procuring concentrating solar cookers and the promotional incentives for box type solar cookers.

---

[25] Ministry of Non-conventional Energy Sources, Annual Report 2000/2001.

[26] Demonstration & Promotion of Solar Photovoltaic Devices/ Systems in Urban Areas,Ministry of new and renewable energy.

- Solar Dyers: The financial support is provided for solar dryers and solar distillation systems for producing pure water. Rural villages are the focus areas.
- Demonstration Scheme for North-East, Islands, J&K and Sikkim : Here the emphasis is on giving financial support for solar water heating systems at sites of high visibility.
- Solar Buildings Programme : In order to promote energy efficient buildng designs based on the principles of solar architecture for rural houses in the villages.[27]

In India the first Solar Thermal Power Plant of 50kW capacity has been installed by MNES following the parabolic trough collector technology at Gwalpahari, Gurgaon, which was commissioned in 1989 and operated till 1990, after which the plant was shut down due to lack of spares.[28] The plant is being revived with various development Components. A Solar Thermal Power Plant of 140MW at Mathania in Rajasthan, has been proposed and sanctioned by the Government in Rajasthan.[29]

SOLAR ENERGY INITIATIVES IN RURAL INDIA

Rural Electrification is on the prime agenda of Government policy. Rural Electrification Policy (2006) envisages the complete electrification of all villages by 2012.[30] The policy proposes that of the 80,000 villages that have no

[27] Solar Thermal Power Generation Programme, Ministry of Renewable Energy, India.

[28] Solar Thermal Tecnologies for Power Generation, available at http://www.geocities.com/pemnq/solarscan.html(last accessed on 28-03-2009).

[29] Shirish Garud, Fellow and Ishan Purohit, 2007. Making solar thermal power generation in India a reality - Overview of technologies, opportunities and challenges, Research Associate. The Energy and Resources Institute (TERI), India.

[30] Rural Electrification Policy2006.

access to electricity, 18,000 villages are in remote areas and that must be electrified through use of renewable energy. Solar energy can play a major role in this. Currently, about 3000 villages have been electrified, primarily through solar systems.[31]

Generally Most of the policies aimed at promoting solar energy was in the form of financial incentives and certain technical and scientific encouragement. For instance, the Ministry of Non-Conventional Energy Sources (MNES) has established test facilities at the Solar Energy Centre, Gurgaon, as well as at six regional test centers in different regions of the country, to facilitate testing of solar collectors for conformity with the national standards.[32] A soft loan scheme has been introduced by the MNES. Soft loans are available at an interest rate of 5% for non-commercial users and 7% for commercial organizations. These loans are available for up to 85% of the cost of the system, and can be repaid over five years.[33] The Government of India announced Special Incentives Package Schemes (SIPS) that apply to solar cell and module manufacturing in the country. Given the nature of incentives specified (capital investment and interest subsidies) there has been enormous response to the announcement in the form of proposals to build cell, wafer and module manufacturing capacity.[34]

The Ministry of New and Renewable Energy has also been encouraging private entrepreneurs and NGOs to set up and operate Solar Shops in major cities of the country with a view to make solar energy products easily available and to provide

---

31 Ministry of New and Renewable Energy, 2007. "Annual Report 2006-07". Government of India.

32 Solar Heat, Ministry of Non-Conventional Energy Sources, Government of India.

33 *Id*

34 The Solar PV Landscape in India, An Industry Perspective,White paper P.V.Group, p3 April 2009.

after sales repair and maintenance services. Finanial incentives to such shops include,

- 5-year loans up to Rs. 1 million (US$ 24,300).
- At 7% interest rate for a maximum of 85% of cost of shop for establishment.
- Monthly Rs. 5,000 ($ 120) grants for the first 2 years.
- Additional monthly Rs. 5,000 grant for the first 2 years based on turnover with 50% to be spent on publicity (see http://mnes.nic.in/swhs-finsupp.htm for details).[35]

INDIA'S ACTION PLAN ON CLIMATE CHANGE ARTICULATES A CENTRAL ROLE FOR SOLAR POWER

The Government of India's National Action Plan on Climate Change released in mid-2008, identifies eight critical missions - one of which is the National Solar Mission. The mission[36], aims at the development of a solar industry competitive with fossil fuels within the next 20-25 years. Specifically, it aims to achieve the goals of 10,000 MW of installed photovoltaics (PV) capacity by 2020.

The NAPCC aims to promote the development and use of solar energy for power generation and other uses with the ultimate objective of making solar competitive with fossil-based energy options. The plan includes:

- Specific goals for increasing use of solar thermal technologies in urban areas, industry, and commercial establishments;
- A goal of increasing production of photovoltaics to 1000 MW/year; and

---

[35] Government of India, Ministry of New and Renewable Energy, "Akshay Urja Shops", accessed August 26, 2008, http://mnes.nic.in/akshayshop.htm.

[36] P P Sangal (2008), India Climate Change Action Plan, *The Economic Times*, Mumbai, July 29, 2008.

- A goal of deploying at least 1000 MW of solar thermal power generation. Other objectives include the establishment of a solar research center, increased international collaboration on technology development, strengthening of domestic manufacturing capacity, and increased government funding and international support.[37]

Though a positive step it is unclear, how the government intends on reaching the 10,000 MW goal of installed PV capacity by 2020, laid out in its 2008 National Solar Mission.

In pursuance of the action plan Government established the solar mission. The mission aims to make India a global leader in solar energy and envisages an installed solar generation capacity of 20,000 MW by 2020, of 1,00,000 MW by 2030 and of 2,00,000 MW by 2050. The mission will operate in three phases. The total expected funding from the government for the 30-year period will run to Rs. 85,000 crore to Rs. 105,000 crore. The first phase will be from(2009-2012) and will emphasize solar deployment intending to achieve rapid scaling-up to drive down costs. It will spur domestic manufacturing through the consolidation and expansion of on-going projects for urban, rural and off-grid applications. The Mission will encourage the use of solar applications to meet day-time peaking power requirement that is now met through diesel generation.[38] In the second phase, to be implemented between 2012 and 2017, the Mission will focus on the commercial deployment of solar thermal power plants. This will involve storage options, and the promotion of solar lighting and heating systems on a large scale in market mode.[39] Finally, between 2017 and 2020, the target is to achieve tariff parity with conventional grid power and achieve an installed capacity of 20 gigawatts (Gw) by 2020. The

[37] National Action Plan on Climate Change, India 2008.

[38] Aarti Dhar, On With a Solar Mission, The Hindu, April 26, 2009.

[39] *Id*

installation of one million rooftop systems with an average capacity of 3 kilowatts (kW) by the same year is also envisaged.[40]

MINISTRY OF RENEWABLE ENERGY GUIDELINES

In terms of policy guidelines the Ministry of New & Renewable Energy (MNRE) has issued guidelines for Generation based incentive for Grid Interactive Solar Photovoltaic power generation projects and Grid Interactive Solar Thermal power generation projects in January 2008 and March 2008 respectively [41]. This is the first elaborate guideline initiated by the ministry regarding solar energy promotion. Salient features of MNRE guidelines providing incentives for Solar PV and Solar Thermal generation projects are:

1. The Ministry will consider support for a maximum capacity up to 50 MW (including solar photovoltaic as well as solar thermal power generation) during the 11th plan period as per the guidelines and norms prescribed.
2. A maximum cumulative capacity of 10MWp of Grid interactive solar PV power and thermal power generation projects of minimum one MW installed capacity will be eligible for consideration of generation based incentive in a State.[42]
3. The Indian Renewable Energy Development Agency (IRDA) will assist the Ministry in fund handling, monitoring and other associated activities in this regard. Details of broad guidelines for submission of proposals by the interested project developers, details

---

[40] *Id*

[41] Guidelines for Generation Based Incentive Grid Interactive Solar PV Power Generation Projects Ministry of New and Renewable Energy,India, and Guidelines for Generation Based Incentive Grid Interactive Solar Thermal Power Generation Projects(http://mnre.gov.in) January, 2008.

[42] *Id*

of incentives and other related guidelines are depicted in the circulars of MNRE.

4. In case any project developer is desirous of availing the accelerated depreciation benefit for the project under section 32 of the Income Tax Act 1961, they would not be eligible for generation-based incentive.[43]

5. The Ministry may provide, through IREDA, a generation-based incentive of a maximum of Rs. 12.00 per kWh (for Solar PV) and Rs. 10.00 (for Solar Thermal) Generation to the eligible projects which are commissioned by 31st December, 2009, after taking in to account the power purchase rate (per kWh) provided by the State Electricity Regulatory Commission or utility for that project. Any project commissioned after 31st December, 2009 would be eligible for a maximum incentive with a 5% reduction and ceiling of Rs. 11.40 per kWh (for Solar PV Generation Project) and Rs. 9.50 (for Solar Thermal Generation Project).[44]

6. The maximum amount of generation based incentive applicable for a project will be determined after deducting the power purchase rate for which PPA has been signed by the utility with a project developer, from a notional amount of Rs 15.00 per kWh for solar PV generation power projects. In all cases the maximum amount of generation-based incentive shall not exceed Rs. 12.00 per kWh.

7. The generation-based incentive will continue to decrease, as and when the utility sign a PPA for power purchase at a higher rate. The proposed annual escalations agreed with the utility, as in force, should also be reflected in the PPA.

---

[43] *Id*

[44] *Id*

8. The generation-based incentive approved for grid interactive Solar PV and Solar Thermal generation projects may be available for a maximum period of ten years from the date of approval.
9. The Ministry may, at any given time, even before 31st March, 2010, announce a new generation based incentive and guidelines, which will be applicable to all such proposals / projects that have not been approved by that time.[45]

Solar Industry responded with a great enthusiasm to the technical and financial incentives given by MNRE. Several Memorandum of Understanding (MoU) were signed by state governments with private companies of magnitude of hundred Mega Watt. But MoU has not been translated into real action. There has not been even a single Solar PV or thermal plant which has been completed or under construction in India. This is because; MNRE guidelines for incentives have been issued without detailed feasibility study and technology identification for solar PV and thermal power plants in India along with proper solar radiation resource assessment.[46] The current status of international technology and its availability and financial and commercial feasibility in the context of India is not clear.

Following specific issues are present in the MNRE guidelines :

- Incentives are only for 10 years whereas the concession period of a solar power plant is usually for 20-25 years. This not only brings the project IRR down but also makes the project unattractive. Although some of the state government like Gujarat has increased the incentive period from 10 to 12 years but it is less to make the solar power plants financially viable.

45 *Id*

46 *Supra* note 30

- Commissioning of power plants has to be before 31 December 2009 to take the maximum incentive under the scheme. This is an ambitious target since most of the equipments need to be imported from outside. Also, Indian companies do not have the skill set for design, built and operate the Solar power plants.
- India does not have a history of solar based power plants. Solar industry will take time to come to terms with the MNRE guidelines affecting different cost parameters.
- Most of the solar plants are proposed to be in barren land. After the Singur and many other Land acquisition incidents, acquiring large pieces of land will be one of the major challenges for a solar based power plant. There should be a discussion on the policy issues concerning land acquisition.
- Cost of producing power from solar panels is higher than the conventional thermal power plants: Conventionally generated electricity ranges between 2 and 4 Rupees per kilowatt hour (the amount of money to get a kilowatt of power for an hour). Solar PV and thermal costs around 10 to 15 Rupees a kilowatt hour. Even with the incentives, it will be tough for the solar companies to run the solar power plants on sustainable basis.

GUJARAT GOVERNMENT - SOLAR POWER POLICY, 2009

Solar power policy announced by Gujarat government is an improvement over the guidelines issued by MNRE. It is worth mentioning here the specific details about the policy.

Solar Power Policy was announced by Government of Gujarat on 7 January. The policy aims at tapping the vast potential for solar energy in the state and transforming it into an "integrated solar generation hub" of the country. The government came up with the solar power policy after it received proposals of up to Rs 50,000 crore in form of MoUs at the Vibrant Gujarat summit. On first day itself, policy was oversubscribed and there were proposals exceeding 500 MW.

- Under the policy, the first of its kind being introduced in Gujarat, the government promises a slew of incentives to prospective developers who may want to set up solar power projects in the state. However, the benefits of this policy will not be available to projects set up under the Union Ministry for Natural and Renewable Energy's (MNRE) incentive scheme for solar power generation. [47]
- Concessions also include exemption from electricity duty and demand cut of 50 per cent of the installed capacity. Units that start production before December 31, 2010 can avail these concessions.
- Under the policy, a maximum of 500 MW project will be allowed for installation, while minimum installed capacity of any project will be five MW.[48]
- Any company or entity will be allowed to set up the solar power project for either captive use or for selling of electricity. However, no fossil fuel will be allowed to be used in the power generation process in the project.
- The power to be generated through the solar energy will be exempted from paying Electricity Duty, while the project developer will have to pay a wheeling charge of two percent of the energy fed to the grid till the wheeling charge is determined by the regulatory body.
- Energy to be generated through solar power will be sold to the distribution licencees in the state at levelised

[47] Solar Power Policy, 2009.

[48] *Id*

fixed tariff per unit for a period of 25 years under a Power Purchase Agreement (PPA).

- As per the tariff mentioned in the policy document for projects commissioned before December 31, 2010, tariff for photo voltaic projects would be Rs 13 per unit for the first 12 years, while subsequently, it would be reduced to Rs 3 from 13th to 25 years.[49]
- For thermal projects to be set up before December 2010, tariff will be Rs 10 per unit for the first 12 years, while it would be reduced to Rs 3 from 13th to 25 years.
- Other projects to be commissioned before March 31 2014, the tariff in the first category (photo voltaic projects) will be Rs 12 per unit for the first 12 years and would be reduced to Rs 3 for the remaining period.[50]

Gujarat and Rajasthan receive more solar irradiation than any other part of the country. So, these states have inherent advantage in developing solar plants and incentives introduced by Gujarat government are definitely an added advantage over other states.

CHALLENGES AND CONSTRAINTS IN SOLAR POWER PROJECT IMPLEMENTATION IN INDIA

Renewable energy technologies (RETs) have long been recognized for their potential as environment-friendly, versatile and sustainable energy alternatives for India. However, despite efforts by the Ministry of Non-conventional Energy Sources (MNES) and the India Renewable Energy Development Agency (IREDA), RETs have not yet succeeded as a major alternative source of energy in rural India. The programmes of MNES and IREDA designed to support small-scale distributed systems have relied on heavily subsidized credit, technology training and consumer awareness activities

---

49 *Id*

50 *Id*

to stimulate the market for renewable energy systems including solar. While there has been some improvements in the field of solar energy in rural areas heavy subsidies have tied the success of these programmes to government budgets and political cycles limiting both the breadth and depth of development/penetration of projects that harness renewable energy resources.[51]

Solar energy is facing some fundamental challenges. Firstly at the policy level different states are having diverse policies which undergo frequent changes leaving the investors and users of renewable energy confused. Need of the hour is to enact a comprehensive national solar policy. Further barriers exist in the form of high cost, its manufacturing procedure as well as its waste products that have some impact on the environment. A lack of necessary transmission infrastructure and high up-front development costs are significant hurdles to developing solar projects on a large scale. Many of the renewable energy technologies are still relatively new to the market, so the commercial chains, networks, marketing and financial links, and other institutional structures that service traditional energy technologies are not in place to assist the entrepreneur even if she/he has the skills, know-how and capital.[52]

Obstacles remain to bringing solar energy costs in line with those of other sources of power generation. The search for better, cheaper solar products are gaining momentum in India. Solar PV now cost one tenth of what they did in early 1980s. Despite the fact that the price of solar photovoltaic technology has been coming down over the years it still remains economically unviable for power generation purposes

---

[51] Commercializing renewable energy in India, Case Studies of Market Transformation: Energy Efficiency and Renewable Energy available at www.un.org / esa / sustdev/publications/energy_casestudies/section4.pdf (Last accessed on 5-04-2009).

[52] Commercializing renewable energy in India, Case Studies of Market Transformation: Energy Efficiency and Renewable Energy, p48,2001

and beyond the reach of most people. At present the initial cost of both types of solar energy systems is higher compared to the cost of conventional energy systems and also the other non-conventional energy systems. With present level of technology, solar electricity produced through the photovoltaic conversion route is 4-5 times costlier than the electricity obtained from conventional fossil fuels. It will be impossible for solar energy to replace fossil fuel unless it has been made accessible at a cheaper rate. Thus providing solar energy at cheaper rate remains the biggest barrier in promoting solar

CONCLUSION

The chapter discussed the various scientific, environmental, legal and policy aspects of Solar energy with regard to India and analyzed its potential to replace fossil fuels. The discussion also revolved on how far our institutional and policy initiatives are tuned to realise our solar potential.

Even though energy from renewable energy sources including solar is growing rapidly, with markets such as solar cells, heaters, cookers experiencing annual double digit growth, the overall solar share is only expected to increase marginally over the coming decades as the demand for energy also grows rapidly, particularly in India. In India, though the emphasis is being given to renewables, some of the better concentration is deliberately moving towards transforming coal into clean energy as well as harnessing hydropower. The recent nuclear deal also to an extent may take away the attention away from renewable.

It is expected that through National Solar mission and other incentives, solar energy production will be given a boost specially the decentralized energy production as it is cost effective. But certain actions of the government suggest a negative response in favour of solar. For instance it was India.

Initially announced that it will set aside US $50 million for subsidizing solar power till its costs would match those of small hydro25, but then allocated them to fund 50 MW of PV by 2012 through a feed-in tariff. While the government has taken a step toward supporting solar through the MNRE feed-in tariff, the policy seems to have been designed as a demonstration programme more than a bona fide support measure. It is restricted to large installations, imposes a very low programme cap, and is funded directly though the government's budget.[53] While there were complaints about the feed-in not be available in conjunction with the tax credit, which had supposedly caused major projects to be reconsidered, the programme was a huge success: barely four months after the publication of the feed-in, 1000 had MW of applications already been submitted.[54] In October 2008, the ministry of new and renewable energy (MNRE) announced its plan to increase the 50 MW cap. It is unclear what the new cap will be, but is a good sign that the government is committing to its Solar Mission to install 10 GW of PV by 2020.

Various government announcements and policy measures suggest that solar energy adoption may be entering a phase of major expansion. But in order to realise the solar potential there is a need of a comprehensive policy and specially subsidy policy for Solar Power in India. Further it is imperative that any such policy for solving energy crisis must cater to environmental concerns and principles of sustainable development.

---

[53] William Hirshman, "On-Grid PV calling?", Photon International, February 2008.

[54] CR Bhattacharjee, "Feed-in-tariff and solar energy", The Financial Express, June 03, 2008 available at (http://www.financialexpress.com/news/Feedintariff-and-solar-energy/317634/)(last accessed on 3-04-2009).

5

Solid Waste Management

A Challenge to Health, Environment and Climate Change

INTRODUCTION

The world is becoming increasingly urban. Historically, urbanization has coincided with, and has been accompanied by, increased consumption and ecological degradation across the globe.[1] The ecological impact of urbanization has not been very positive. Some of the obvious consequences include the problem of depletion of resources, the all pervading pollution, climate change and mismanagement of solid waste. This negative fallout has become a major justification for a new development paradigm: that of sustainable development a development strategy which meets the needs of the present generations without compromising the needs of the future generations. In the environment policy rhetoric the concept of sustainable development with its offshoots principles like polluter pays, precautionary rule, public trust doctrine and intergenerational equity became the central theme. In connection with urbanization and its pursuit of achieving sustainable development concept of sustainable cities is gaining policy attention. Framing an effective Solid waste management assumes significance in the pursuit of

[1] Darshini Mahadevial, "Sustainable Urban Development In India: An Inclusive Perspective", Development in Practice. May 11(2-3):242-59,(2001).

sustainable cities. The purpose of the chapter is to analyse the efficacy of solid waste management in mitigating climate change. The chapter explores the strengths and weaknesses of the policy measures aimed at mitigating climate change and makes appropriate recommendations for its improvement.

URBANIZATION AND MUNICIPAL SOLID WASTE GENERATION

The world urban population is expected nearly to double by 2050, increasing from 3.3 billion in 2007 to 6.4 billion in 2050. By mid-century the world urban population will likely be the same size as the world's total population in 2004. It is expected that during 2008, for the first time in history, the proportion of the population living in urban areas will reach 50 per cent[2] (Fig. 5·1).

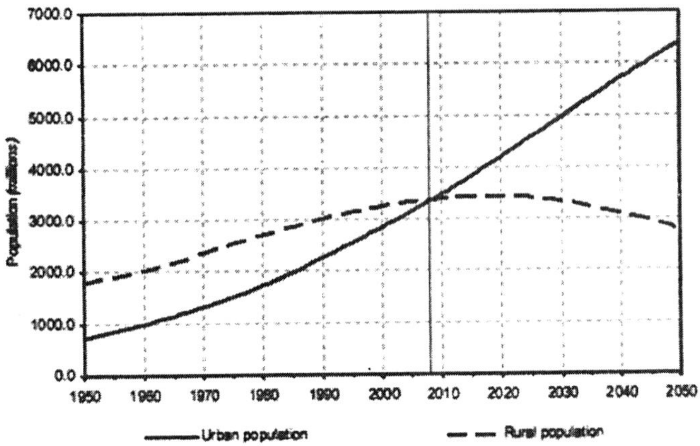

Figure 5·1: Urban and Rural Populations of the World, 1950-2050

Source: United Nations 2008

While in the more developed regions, the proportion urban was already nearly 53 per cent in 1950, in the less

2 World Urbanization Prospects. The 2007 Revision, Executive Summary, Economic and Social Affairs, United Nations,2008.

developed regions the 50 per cent level will likely be reached around 2019 (Fig. 5·2).

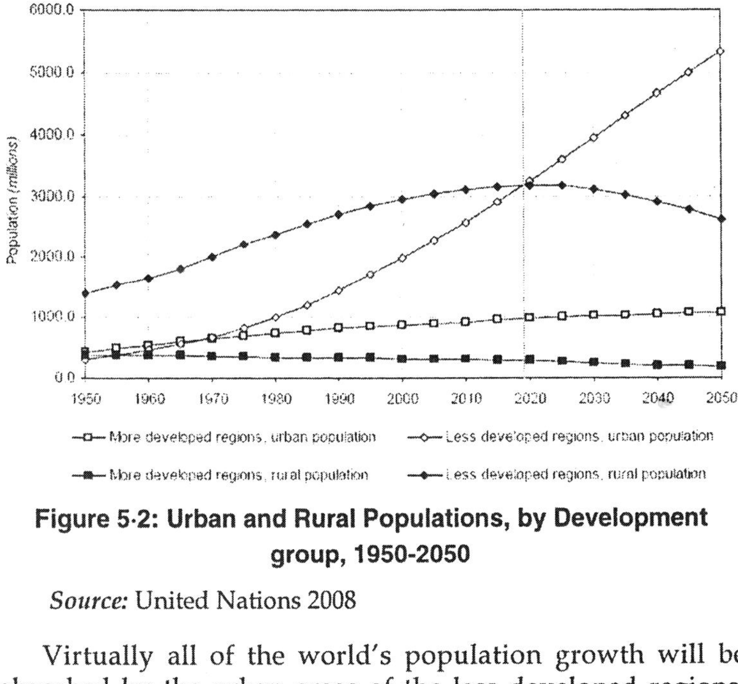

Figure 5·2: Urban and Rural Populations, by Development group, 1950-2050

Source: United Nations 2008

Virtually all of the world's population growth will be absorbed by the urban areas of the less developed regions, whose population is projected to increase from 2.4 billion in 2007 to 5.3 billion in 2050. The urban population of the more developed regions is projected to increase modestly, from 0.9 billion in 2007 to 1.1 billion in 2050.

As an offshoot of urbanization and rapid growth of population the generation of Municipal Solid waste (MSW) has seen substantial rise in the last few decades. The population growth directly contributes to the growth in solid waste generation in urban areas putting pressure on the existing infrastructure.[3] For instance, following table highlights the corresponding increase in generation of solid waste with increase in population in Delhi, India. The table

---

[3] Arlosoroff, S. " Developing Countries Struggle with Waste Management Policies', *Waste Management and Research 9*: 491-494,(1991).

suggests with every increase in population total waste generation as well as per capita waste generation increased.

Table 5·0: Solid Waste Generation in Delhi

Year	*2010*	*2015*	*2020*	*2024*
Population	1.77 Crore	1.99 Crore	2.23 Crore	2.45 Crore
Daily Visitors	0.07 Crore	0.078 Crore	0.087 Crore	0.095 Crore
Per Capita Waste Generation	500 Gms/day	700 Gms/day	725 Gms/day	750 Gms/day
Waste Generation	9200 (TPD)	14475(TPD)	16800(TPD)	19100 (TPD)

Source : **Delhi Environment Status Report, WWF, 2005**

SOLID WASTE, COMPOSITION, HAZARDS AND CLIMATE CHANGE

Solid waste refers to all non liquid wastes. Municipal solid waste refers to the total waste generated in municipalities. Though the composition and type of waste generated varies from place to place certain types of solid wastes are commonly generated (Table 5·1).

Table 5·1: Source and Type of Solid Waste

Source	*Typical waste generators*	*Type of waste*
Residential	Single and multifamily dwellings.	Food wastes, paper, cardboard, plastics, textiles, leather, yard wastes, wood, glass, metals, ashes, special wastes (e.g., bulky items, consumer electronics, white goods, batteries, oil, tires, and household hazardous wastes).
Industrial	Light and heavy manufacturing, fabri cation, construction sites, power and chemical plants.	Housekeeping wastes, packaging, food wastes, construction and demolition materials, hazardous wastes, ashes, special wastes.

Commercial	Stores, hotels, restaurants, markets, office buildings, etc.	Paper, cardboard, plastics, wood, food wastes, glass, metals, special wastes, hazardous wastes.
Institutional	Schools, hospitals, prisons, government centers.	Same as commercial.
Construction and demolition	New construction sites, road repair, renovation sites, demolition of buildings.	Wood, steel, concrete, dirt, etc.
Municipal services	Street cleaning, landscaping, parks, beaches, other recreational areas, water and wastewater treatment plants.	Street sweepings; landscape and tree trimmings; general wastes from parks, beaches, and other recreational areas; sludge.
Agriculture	Crops, orchards, vineyards, dairies, feedlots, farms	Spoiled food wastes, agricultural wastes, hazardous wastes (e.g., pesticides).

Source [4]

Whatever may be the type and composition solid waste poses several major problems. These includes.

- Health and environmental hazards from uncollected waste;
- Health and environmental hazards from collected but poorly disposed of waste;
- The economic burden of waste disposal on towns and cities.[5]

Uncollected waste especially excreta and other liquid and solid waste from households and the community, are a serious health hazard and lead to the spread of infectious diseases.

[4] Hoornweg, Daniel, Laura Thomas," What A Waste: Solid Waste Management in Asia,. Working Paper Series Nr. 1. Urban Development Sector Unit.,East Asia and Pacific Region. Page 5 Table 22,(1999).

[5] David Pearce, "Economics and Solid Waste Management In The Developing World", CSERGE Working Paper WM 94-05.

Unattended waste lying around attracts flies, rats, and other creatures that in turn spread disease. Other hazards include piles of rubbish as impediments to traffic and the general lack of aesthetic appearance of city streets besides creating pollution. Secondly scavengers operating on landfill sites, without proper facilities and equipment, are typically exposed to a range of public health and environmental hazards associated with open landfill sites. During periods of heavy rain stagnant water ponds are commonly found on such sites. These serve as breeding grounds for mosquitoes which are vectors of infective agents spreading malaria and filariasis.[6] The group typically at risk from the open and unscientific dumping of solid waste include - the population in areas where there is no appropriate waste disposal method, especially the pre-school children; waste workers; and workers in facilities producing toxic and infectious material. Other high-risk group include population living close to a waste dump and those, whose water supply has become contaminated either due to waste dumping or leakage from landfill sites.[7]

The various environmental and health are generally known but what most people don't realize that solid waste emissions result in global climate change. Solid waste is now increasingly viewed as an environmental, health and climate change problem. The manufacture, distribution and use of products - as well as the production of waste as well as management of the resulting waste - all result in greenhouse gas emissions.[8] For instance solid waste landfills are the single largest man-made source of methane gas in the United States. Methane (CH_4) is a powerful greenhouse gas that is 23 times

[6] *Id*

[7] Margaret N. Maxey," Hazards of Solid Waste Management.Bioethical Problems, Principles, and Priorities", Environmental Health Perspectives, Vol. 27, pp. 223-230, (1978).

[8] U.S. Environmental Protection Agency, Inventory of U.S. Greenhouse Gas Emissions and Sinks: 1990 - 1997, April 1999 [EPA 236-R-99-003].

more effective at trapping heat in the atmosphere than the most prevalent greenhouse gas, carbon dioxide (CO_2). 36 per cent of human caused methane releases come from our municipal solid waste landfills. Organic materials (derived from living organisms) produce methane in landfills when they decompose without oxygen, under tons of garbage. A ton of municipal solid waste landfilled produces 123 pounds of methane.[9]

Waste prevention and recycling reduce greenhouse gases associated with these activities by reducing methane emissions, saving energy, and increasing forest carbon sequestration.

The U.S. Environmental Protection Agency estimates that by cutting the amount of waste generated back to 1990 levels, US could reduce greenhouse gas emissions by 11.6 million metric tons of carbon equivalent (MTCE), the basic unit of measure for greenhouse gases.[10] Increasing US national recycling rate from its current level of 28 per cent to 35 per cent would reduce greenhouse gas emissions by 9.8 million MTCE,[11] compared to landfilling the same material. Together, these levels of waste prevention and recycling would slash emissions by more than 21.4 million MTCE – an amount equal to the average annual emissions from the electricity consumption of roughly 11 million households. Making a ton of aluminium cans from its virgin source, bauxite, uses 229 British thermal units (Btus). In contrast, producing cans from recycled aluminium uses only 8 Btus per ton, an energy savings of 96%.[12] Extracting and processing petroleum into

---

[9] Climate change and waste, available at http://www.epa.gov/climatechange/wycd/waste/index.html (Last accessed on 2-03-2009).

[10] *Id*

[11] Bill Sheehan,Zero Waste, Recycling and Climate Change,Grass Roots Recycling Network available at http://www.grrn.org/zerowaste/climate_change.html

[12] Jeffrey Morris, "Recycling vs. Incineration: An Energy Conservation Analysis," *Journal of Hazardous Materials* 47, pp 277-293,(1996).

common plastic containers (No. 1 'PET' and No. 2 'HDPE') takes four to eight times more energy than making plastics from recycled plastics.[13] This paper analyses how a proper solid waste management system can promote the growth of sustainable cities and help in curbing the green house gases (GHG) emissions.

SOLID WASTE GENERATION IN INDIA: CURRENT ESTIMATES

With industrial progress, growing urban areas and resultant growth in urban solid wastes is a relatively new phenomenon in contemporary India. Total quantity of waste generated in the country based on weight meant exercise by local bodies is not reported. However, Ministry of Urban Development in its manual on solid waste management (year 2000) has estimated waste generation of 100,000 MT.[14]

Table 5·2: Quantities and Waste Generation Rate in 59 Cities is as Under.

S. No	*Name of City*	*Population (As per 2001 census)*	*Area (Sq. Km)*	*Waste Quantity (TPD)*	*Waste Generation Rate (kg/c/day)*
1	Kavaratti	10,119	4	3	0.30
2	Gangtok	29,354	15	13	0.44
3	Itanagar	35,022	22	12	0.34
4	Daman	35,770	7	15	0.42
5	Silvassa	50,463	17	16	0.32
6	Panjim	59,066	69	32	0.54
7	Kohima	77,030	30	13	0.17
8	Port Blair	99,984	18	76	0.76
9	Shillong	1,32,867	10	45	0.34
10	Shimla	1,42,555	20	39	0.27

[13] Brenda Platt and Neil Seldman, "Wasting and Recycling in the United States 2000, *Prepared for* GRRN, Institute for local Governance.

[14] Central Pollution Control Board, Waste.

11	Agartala	1,89,998	63	77	0.40
12	Gandhinagar	1,95,985	57	44	0.22
13	Dhanbad	1,99,258	24	77	0.39
14	Pondicherry	2,20,865	19	130	0.59
15	Imphal	2,21,492	34	43	0.19
16	Aizwal	2,28,280	117	57	0.25
17	Jammu	3,69,959	102	215	0.58
18	Dehradun	4,26,674	67	131	0.31
19	Asansol	4,75,439	127	207	0.44
20	Kochi	5,95,575	98	400	0.67
21	Raipur	6,05,747	56	184	0.30
22	Bhubaneswar	6,48,032	135	234	0.36
23	Tiruvanantapuram	7,44,983	142	171	0.23
24	Chandigarh	8,08,515	114	326	0.40
25	Guwahati	8,09,895	218	166	0.20
26	Ranchi	8,47,093	224	208	0.25
27	Vijaywada	8,51,282	58	374	0.44
28	Srinagar	8,98,440	341	428	0.48
29	Madurai	9,28,868	52	275	0.30
30	Coimbatore	9,30,882	107	530	0.57
31	Jabalpur	9,32,484	134	216	0.23
32	Amritsar	9,66,862	77	438	0.45
33	Rajkot	9,67,476	105	207	0.21
34	Allahabad	9,75,393	71	509	0.52
35	Vishakhapatnam	9,82,904	110	584	0.59
36	Faridabad	10,55,938	216	448	0.42
37	Meerut	10,68,772	142	490	0.46
38	Nashik	10,77,236	269	200	0.19
39	Varanasi	10,91,918	80	425	0.39
40	Jamshedpur	11,04,713	64	338	0.31
41	Agra	12,75,135	140	654	0.51
42	Vadodara	13,06,227	240	357	0.27
43	Patna	13,66,444	107	511	0.37
44	Ludhiana	13,98,467	159	735	0.53
45	Bhopal	14,37,354	286	574	0.40
46	Indore	14,74,968	130	557	0.38

47	Nagpur	20,52,066	218	504	0.25
48	Lucknow	21,85,927	310	475	0.22
49	Jaipur	23,22,575	518	904	0.39
50	Surat	24,33,835	112	1000	0.41
51	Pune	25,38,473	244	1175	0.46
52	Kanpur	25,51,337	267	1100	0.43
53	Ahmedabad	35,20,085	191	1302	0.37
54	Hyderabad	38,43,585	169	2187	0.57
55	Banglore	43,01,326	226	1669	0.39
56	Chennai	43,43,645	174	3036	0.62
57	Kolkata	45,72,876	187	2653	0.58
58	Delhi	1,03,06,452	1483	5922	0.57
59	Greater Mumbai	1,19,78,450	437	5320	0.45

Source: CPCB Data

In India data is clear there is a positive correlation between population and the generation of waste. Generally increased population is accompanied by increased waste generation. Data on the per capita MSW reveals that it normally vary from 0.17 to 0.76 Kg/c/day.But the data on the per capita MSW quantity from Subban studies show that the per capita quantity varies from 0.21kg/capita/day for small cities (0.1-0.5 million) to 0.50 kg/capita/day for big cities (>2.0 million).[15] Large towns show distinctly higher per capita waste generation, while there is no significant difference between medium and small towns.[16]

The composition of the MSW is directly influenced by the food habits, culture, and socio-economic and climatic conditions. Physical and chemical characteristics indicate that NEERI has extensively characterized solid waste for 43 cities and showed the organic content of the samples on a dry weight basis to vary between 20 and 40% (Table 5·3) with

[15] Subban R., Strategy Paper on "Solid Waste Management in India", Water supply and sanitation in urban environment in India. NEERI, Nagpur, India, (1996).

[16] *Id*

nitrogen, phosphorus and potassium content of the MSW ranging between 0.5-0.7%, 0.5-0.87% and 0.5-0.8% respectively. Factors influencing the quantities and composition include the average level of income, the sources, the population, social behaviour, climate, industrial production and the existence of markets for waste materials. Waste densities and moisture contents are much higher in India, which require different technology and management systems.[17]

Table 5·3: Composition of MSW for 55 Cities

S. No	*Name of City*	*Compostables (%)*	*Recyclables (%)*	*C/N Ratio*	*HCV* (Kcal/ Kg)*	*Moisture (%)*
1	Kavarati	46.01	27.20	18.04	2242	25
2	Gangtok	46.52	16.48	25.61	1234	44
3	Itanagar	52.02	20.57	17.68	3414	50
4	Daman	29.60	22.02	22.34	2588	53
5	Silvassa	71.67	13.97	35.24	1281	42
6	Panjim	61.75	17.44	23.77	2211	47
7	Kohima	57.48	22.67	30.87	2844	65
8	Port Blair	48.25	27.66	35.88	1474	63
9	Shillong	62.54	17.27	28.86	2736	63
10	Shimla	4302	3664	2376	2572	60
11	Agartala	58.57	13.68	30.02	2427	60
12	Gandhinagar	34.30	13.20	36.05	698	24
13	Dhanbad	46.93	16.16	18.22	591	50
14	Pondicherry	49.96	24.29	36.86	1846	54
15	Imphal	60.00	18.51	22.34	3766	40
16	Aizwal	54.24	20.97	27.45	3766	43
17	Jammu	51.51	21.08	26.79	1782	40
18	Dehradun	51.37	19.58	25.90	2445	60
19	Asansol	50.33	14.21	14.08	1156	54
20	Kochi	57.34	19.36	18.22	591	50

[17] Cointreau, S., Gunnerson, C.G., Huls, J.M. and Seldman, N., "Recycling from Municipal Refuse: A State-of-the-Art Review and Annotated Bibliography". *UNDP Project Management Report* No. 1, World Bank Technical Paper No. 30, World Bank, Washington, D.C. (1984).

21	Raipur	51.40	16.31	223.50	1273	29
22	Bhubaneswar	49.81	12.69	20.57	742	59
23	Tiruvananthapuram	72.96	14.36	35.19	2378	60
24	Chandigarh	57.18	10.91	20.52	1408	64
25	Guwahati	53.69	23.28	17.71	1519	61
26	Ranchi	51.49	9.86	20.23	1060	49
27	Vijaywada	59.43	17.40	33.90	1910	46
28	Srinagar	6177	17.76	22.46	1264	61
29	Madurai	55.32	17.25	32.69	1813	46
30	Coimbatore	50.06	15.52	45.83	2381	54
31	Jabalpur	58.07	16.61	28.22	2051	35
32	Amritsar	65.02	13.94	30.69	1836	61
33	Rajkot	41.50	11.20	52.56	687	17
34	Allahabad	35.49	19.22	19.00	1180	18
35	Visakhapatnam	45.96	24.20	41.70	1602	53
36	Faridabad	42.06	23.31	18.58	1319	34
37	Meerut	54.54	10.96	19.24	1089	32
38	Nasik	39.52	25.11	37.20	2762	62
39	Varanasi	45.18	17.23	19.40	804	44
40	Jamshedpur	43.36	15.69	19.69	1009	48
41	Agra	46.38	15.79	21.56	520	28
42	Vadodara	47.43	14.50	40.34	1781	25
43	Patna	51.96	12.57	18.62	819	36
44	Ludhiana	49.80	19.32	52.17	2559	65
45	Bhopal	52.44	22.33	21.58	1421	43
46	Indore	48.97	12.57	29.30	1437	31
47	Nagpur	47.41	15.53	26.37	2632	41
48	Lucknow	47.41	15.53	21.41	1557	60
49	Jaipur	45.50	12.10	43.29	834	21
50	Surat	56.87	11.21	42.16	990	51
51	Pune	62.44	16.66	35.54	2531	63
52	Kanpur	47.52	11.93	27.64	1571	46
53	Ahmedabad	40.81	11.65	29.64	1180	32
54	Hyderabad	54.20	21.60	25.90	1969	46
55	Bangalore	5184	2243	3512	2386	55

Source: CPCB Data

SOLID WASTE MANAGEMENT: PROMISES AND PITFALLS

The major key components of solid waste management consists of,

- Generation
- Storage
- Collection
- Transport
- Disposal

Several problems including technical, financial and institutional constraints plague all the five stages of municipal solid waste management. Solid waste management account for 20% to 50% of available operational budgets for municipal services. Gross inefficiencies are quite common in their functioning. Even with a substantial budget support they provide a service for 50% of the urban population and collecting only 60% to 70% of the refuse arisings.[18] Following table illustrate the low amount of spending on solid waste management in low income countries which adds to the existing environment and hygienic problems.

Table 5·4: Municipal Waste Arisings and Disposal in Developing Countries[19]

	Low Income	*Middle Income*	*High Income*
	\$350 p.c.p.a.	\$1950 p.c.p.a.	\$17,500 p.c.p.a.
Arisings per capita	200 kg p.c.p.a.	300 kg p.c.p.a.	600 kg p.c.p.a.
Waste per \$ income	0.51 kg p. \$	0.15 kg p. \$	0.03 kg p. \$

Source: Cointreau-Levine (1992).

[18] Bartone, C., Leite, L., Triche, T. and Schertenleib, R., "Private Sector Participation in Municipal Solid Waste Service: Experiences in Latin America". *Waste Management and Research* 9: 495-509,(1991).

[19] Cointreau-Levine, S.J. "Private Sector Participation in Municipal Solid Waste Services in Developing Countries". *Paper for the Infrastructure and Urban Development Department,* World Bank, Washington DC,(1992).

GENERATION

The solid waste quantities generated in urban centers are increasing due to rise in the population and increase in the per capita waste generation rate.[20] On the basis of available data, it is estimated that the nine major metropolitan centres in India are presently producing 23,000 tonnes of solid waste per day. While about 50,000 tonnes are generated daily from the Class I cities.

COLLECTION AND TRANSPORT

Piles of Garbage and wastes of all kinds littered everywhere have become common sight in our urban life. Study conducted in 11 major African cities waste collection systems showed that 30 to 50% of the generated solid waste is not collected.[21] Pearce estimated that as much as 60 MT of wastes per annum may remain uncollected in low-income countries and a further 45 MT in middle-income countries.[22] Inefficient collection technique exacerbates this problem. Most of the garbage is taken to the disposal sites by means of uncovered vehicles. It has been estimated that the cost of cleaning these MSW off streets in developing countries is two to three times that of actual collection cost. Regarding India the same story continues due to budgetary constraints, inadequate equipment and poor planning, house- to- house collection is very rare in India, particularly in certain low-

[20] Baldisimo, J.M. (ed)., "Scavenging of Municipal Solid Waste in Bangkok, Jakarta, and Manila." *Environmental Sanitation Reviews*. December no.26. Asian Institute of Technology, Bangkok,(1988).

[21] Aarekezi S., "Renewable Energy Technologies as an Option for a Low-Carbon Energy Future for Developing Countries: Case Examples from Eastern and Southern Africa. *African Energy Policy Research Network* (AFREPREN). Foundation for Woodstove Dissemination (FWD) Nairobi, Kenya.

[22] Pearce D. and R. K. Turner, "Economics and Solid Waste Management in the Developing World", *CSERGE Working Paper* No. WM- 9404,(1982).

income areas where waste is not collected at all.[23] It is estimated that up to 30 per cent of disposed solid waste (public dustbins and street garbage) are left uncollected. The areas which are not serviced are left with clogged sewers and litter which creates serious health problems for the resident population.[24] Bangalore, the capital of the southern state Karnataka, with an estimated population of 5 million and one of the largest cities of India, is estimated to generate around 3000 to 3600 tonnes of municipal waste per day, out of which 80-85 per cent is collected, and the rest left unattended.

DISPOSAL

The prevalent methods of solid waste disposal are the uncontrolled dumping and burning on open ground and city streets.[25] The problem plaguing India and other developing countires is that 80-90% of the budget spending is on collection and transportation and very less is spending on treatment. In developing countries SWM is centralized concept and therefore 85-90% of the municipal SWM budget goes in waste collection and transportation and only 0.8 % on treatment technologies. Decentralization of the treatment and selection of effective treatment options would enhance the efficiency and reduce the cost of treatment. A detailed assessment of the three key energy-from-waste options (incineration, anaerobic digestion and land-filling) using 4 parameters related to waste composition, quantity of waste required, capital cost, and land requirement showed that landfilling should be given priority over the other two options

---

[23] Baud I.S.A, Schenk H, New Approaches to Urban Solid Waste Management: linkages between formal and informal systems of source separation, collection and recycling in Indian cities,publication of the University of Amsterdam(1994).

[24] NIUA, Handbook of Urban Statistics, 1993.

[25] Cointreau-Levine, J. Sandra, "Private sector Participation in Municipal Solid Waste Services in Developing Countries": *Volume 1 of The Formal Sector*, Washington DC; World Bank,(1994).

because of its low-cost, comparatively modest investment requirement and relatively simpler equipment that does not require sophisticated operation and maintenance skills.[26] Cointreau-Levine (1992) has suggested some benchmark figure for the costs of disposal in developing countries (Table 5·5).

Table 5·5: Benchmark Costs for Municipal Waste Collection and Disposal ($ per tonne)[27]

Cost Items	*Low Income*	*Medium Income*	*High Income*
Collection	15-30	30-70	70-120
Disposal	1-3	3-10	15-50
Transfer	3-5	5-15	15-20
Total (Maximum)	19-38	38-95	100-190

The technological options for SWM are incineration, anaerobic digestion, composting. Like other developing Asian countries, in India too most of the waste is landfilled. The wastes are largely dumped. This dumping is normally carried out in low-lying areas, which are prone to flooding and leads to ground water pollution. A consequence of this waste generation and disposal is a rise in methane emissions, which contribute to global warming. Annual emissions of methane from landfills were estimated to be about 7 million tons in 1997, and are expected to rise to 39 million tons by 2047.[28] There are specific concerns about the health effects of emissions from landfill on local communities. Several studies have been carried out.[29] Given that the research in this area is

[26] Green Paper on the management of bio-waste in the European Union, SEC 2008.

[27] *Supra* note Cointreau-Levine 19.

[28] Singhal, S. and S Pandey, Solid waste management in India: status and future directions. TERI Information Monitor on Environmental Science, Vol. 6, No. 1, pp 1-4.,(2001) available at http://www.teriin.org/envis/times6-1.pdf(Last accessed on 3-03-2009).

[29] Birth Outcomes and Selected Cancers in Populations Living Near to Landfill Sites., Report to the Department of Health, Aug 2001.,Paul Elliott et

Contd....

at an early stage, any connection between changing climate conditions and the health impacts of emissions from landfill would only be speculative.

A recent report attempted to estimate the disamenity costs and externalities of landfill. Disamenity encapsulates the nuisance caused locally by the landfill and includes noise, dust, litter, odour and vermin.[30] The report concluded that there was a reduction in house price near to landfill sites. Some of this was explained by physical and socio-economic factors but taking these into consideration, there was still a statistically significant stock disamenity effect for houses closer than 0.5 miles from a landfill. This is one of the main problems in the location of landfill sites in urban areas in India. Local people protest in development of a landfill in their neighbourhood.

MUNICIPAL AGENCIES SPEND ABOUT 5-25% OF THEIR BUDGET ON MSWM.[31]

It is broadly estimated that between 10 to 40 per cent of the total municipal budget is used for SWM (Bhide, 1990). However, despite this large share, it is argued that the Indian waste management system is starved of resources when account is taken of the increasing demands associated with growing urbanisation (Shekdar, et. al., 1992). Data availability is very less and not very reliable.In spite of such heavy

---

Contd. from previous page...

al, The Small Area Health Statistics Unit (SAHSU), Dept. of Epidemiology and Public Health, imperial college St Mary's Campus, Norfolk Place, London W2 1PG. Available at HYPERLINK "http://www.doh.gov.uk/landh.htm" www.doh.gov.uk/landh.htm(last accessed on 3-04-2009).

[30] Jonatan Bebb, Jim Carcey "Potential impact of Climate Change on Waste Management"Environment Agency,U.K,(2003) .

[31] Gurdeep Singh, Tauseef Zia Siddiqui and Anshul Jain, Sustainable Development through Integrated Municipal Solid Waste Management (MSWM) Approach - A Case Study of Indian School of Mines Campus, Proceedings of the International Conference on Sustainable Solid Waste Management,5 - 7 September 2007, Chennai, India. pp.54-58.

expenditure, Management of Municipal Solid Wastes (MSW) continues to remain one of the most neglected areas of urban development in India. The present status of MSW situation is so grave that in many urban areas there is a real threat to the public health in particular and the environmental quality in general.

The increasing solid waste quantities and the areas to be served strain the existing SWM system. The equipment and machinery presently used in the system are usually that which have been developed for general purpose or that which have been adopted from other industry. Mostly out of the total expenditure, around 90% is accounted for manpower of which major portion is utilized for collection. The operational efficiency of SWM depends on the active participation of both the municipal agency and the citizens. The SWM system is unplanned and is operated in an unscientific way. Neither the work norms are specified nor the work of collection staff appropriately supervised. The vehicles are poorly maintained and no schedule is observed for preventive maintenance. Due to shortage of financial resources, the vehicles are often used beyond their economical life resulting in inefficient operation.[32]

LEGAL AND INSTITUTIONAL FRAMEWORK IN INDIA

In India, solid waste management services are provided by the Civic Bodies as per the provisions of the respective Corporation/Municipal/Panchayat Acts. Many of the Acts are quite old and the provisions need amendments to reflect the changes in the waste management needs. Except for the metropolitan cities where a separate department headed by an engineer exists, the health officer is entrusted with the responsibility of SWM along with other activities.

---

[32] Sunil Kumar, "Municipal Solid Waste Management in India: Present Practices and Future Challenge", available at http://www.cd3wd.com/cd3wd_40/ASDB_SMARTSAN/Kumar.pdf(Last accessed on 5-03-2009).

An analysis of legal provisions brings to our notice that initially there was no explicit statute dealing with solid waste management. The issues relating to solid waste were scattered in provisions of Indian Penal Code, Criminal Procedure Code and Constitution. Under Constitution we can read solid waste management in the Non-justifiable provisions of Directive principles and Fundamental duties. Article 47 of the Indian Constitute makes it a paramount principle of governance that 'steps are taken for the improvement of public health as amongst its primary duties.'[33] Further Art 51A (g) points out that it is the duty of individual to protect and improve the natural environment including forests, lakes, rivers and wild life, and to have compassion for living creatures.[34]

The Indian Penal Code of 1860 has tried to address the issue of solid waste management under offences affecting the public health, safety, convenience, decency and morals'.[35] Since, solid waste can seriously affect the health of the community by giving birth to various diseases, the issue of solid waste has been treated as 'public nuisance' and has been made punishable.[36] But there is no direct section in the Code which deals with the problem of solid waste. Section 133 of the Criminal Procedure Code, 1973 provides a speedy and summary remedy in case of urgency where damages to public interest or public health etc. are concerned. The section deals with 'removal of nuisance' and empowers the Sub-Divisional Magistrate or any executive Magistrate, on receiving report/information, to make order to remove the public

[33] Constitution of India Article 47.

[34] *Id*, Art 51A.

[35] Indian Penal Code 1860 Section 268 to278.

[36] *Id*, Section 268 provides, A person is guilty of a public nuisance who does not act or is guilty of an illegal omission which causes any common injury, danger or annoyance to the public or to the people in general who dwell or occupy property in the vicinity, or which must necessarily cause injury, obstruction, danger or annoyance to persons who may have occasion to use any public right. A common nuisance is not excused on the ground that it causes some convenience or advantage.

nuisance and desist from carrying any trade, business which is causing public nuisance.[37] In fact the section bore the seed of judicial activism in environmental jurisprudence and courts have frequently interpreted and used the section so deal with issue of nuisance. In the famous case of *Municipal Corporation, Ratlam* v. *Shri Vardhichand*[38] Justice Krishna Iyer declared that '...the guns of Section 133 go into action wherever there is public nuisance. The public power of the Magistrate under the Code is a public duty to the members of the public who are victims of the nuisance.'[39] If the order is defied or ignored, Section 188, I.P.C. comes into penal play. It was further, observed that 'imperative tone of S. 133, Cr.P.C. read with the punitive temper of S. 188 I.P.C. makes the prohibitory act a mandatory duty.'

Taking a clue from the positive interpretation of section 133 and the widest expanding interpretation of article 21 of the constitution the highest judiciary on various cases proclaimed unequivocally that right to a clean and wholesome environment comes within the purview of Art 21 of the constitution.[40] The court have also declared that it is a primary, mandatory and obligatory duty of the municipal corporations/ councils to remove rubbish, filth, night soil or any noxious or offensive matter. The Pollution Boards and its officers have a basic duty under the Environment (Protection) Act, 1986 to stop unauthorised movement and/or disposal of the waste. They are also empowered to take action against erring industries and persons. In *CMC Vellore* v. *State of Tamilnadu* [41] and in many other cases, the Supreme Court has time and again declared that right to life under Article 21 encompasses right to live with human dignity, quality of life,

---

37 Criminal Procedure Code 1973,Section 133.

38 Municipal Council, Ratlam v. *Vardichand*, (1980) 4 SCC 162.

39 *Id*

40 *Chetriya Pardushan Mukti Sangarsh Samiti* v. *State of UP* (AIR 1990 SC 2060).

41 AIR 1996 Sc 2715.

and decent environment. Thus, pollution free environment and proper sanitary condition in cities and towns, without which life cannot be enjoyed, is an integral facet of right to life. Highest judiciary paid its specific attention to solid waste management in the case of *B.L Wadhera* v.*Union Of India*.[42] In the case the petitioner approached the court for a direction to the MCD and the NDMC to perform their statutory duties in particular the collection, removal and disposal of garbage and other waste. The Court also issued certain directions.

The complexity of the situation unfolded, as the authorities did not act as per the directions of the court. A sequel followed to the above case and Almitra Patel series of cases in respect of solid waste management followed.[43] In the first case of the series the Supreme Court had to form a committee to examine among other issues "To review municipal bye-laws and the powers of local bodies and regional.

The court directed the Municipal Commissioner that in relation to sanitation and public health, prohibit accumulation of any rubbish, filth, garbage or other polluted obnoxious matters in any premises and/or prohibiting any person from depositing the same in any street or public place shall be scrupulously complied. Court further directed that the streets, public premises such as parks etc. shall be surfaced cleaned on daily basis, including on sundays and public holidays.[44] Direction also included levying and recovering charges and costs from any person littering or violating provisions of the diverse Acts, bye-laws and Regulations relating to sanitation and health for violating the directions being issued herein. For this wide publicity needs to be given, to ensure proper and scientific disposal of waste in a manner so as to sub serve the common good. In this connection they shall endeavour to

42 AIR1996 Sc 2969

43 *Almitra H. Patel* v. *Union of India*, 1998-(002)-SCC -0416.

44 *Almitra H. Patel* V. *Union Of India* 2000-(002)-SCC -0166.

comply with the suggestions and directions contained in the report prepared by the Asim Burmon Committee. Further they have been directed, identify and make available to the MCD and NDMC within four weeks from today sites for setting up compost plants.[45]

But after nearly four years from B.L.Wadehra and two years from Almitra, when the matter came up before the court, the Supreme Court was faced with the situation of inaction by the authorities and in exasperation observed that "We believe it is not for this Court to direct as to how the municipal authorities should carry out their functions and resolve difficulties in regard to the management of solid waste. The Court, in fact, is ill equipped to do so. Without doubt the governmental agencies including the local authorities have all the powers of the State to take action and ensure that the city remains clean. They have only to wake up and act."[46]

The Municipal Solid Waste (Management and Handling) Rules, 2000[47] is the first direct legislative attempt at tackling the solid waste problem. These rules consist of certain concrete guidelines aimed at making the handling of solid waste more sustainable and efficient. Rule 4 directs the municipal authorities to develop infrastructure development for collection, storage, segregation, transportation, processing and disposal of municipal solid wastes. It also imposes duty on the municipal authorities to submit an annual report on implementation to Ministry of Urban development.

COLLECTION ON MUNICIPAL SOLID WASTE

The rule prohibits the littering of MSW in cities, towns etc. It plans to collect waste on regular basis from certain areas which are normally the ground of solid waste like slums,

---

[45] *Id*

[46] *Almitra H. Patel* v. *Union of India* 2000-(002)-SCC -0679.

[47] The Municipal Solid Wastes (Management and Handling) Rules, 2000.

squatter areas, slaughter house, fish markets, fruits and vegetable, industrial waste, horticultural and construction or demolition waste etc. Another significant feature of the rules relates to the organised house-to-house collection of waste. Further the rules impose the responsibility of the generator of waste to avoid littering and delivery of waste in accordance with collection and segregation system.[48]

SEGREGATION OF MUNICIPAL SOLID WASTES

Non-segregation of waste poses a major handicap in effectively treating the solid waste occurs. Encouraging Community participation was considered to be essential in segregation of waste. It was provided in order to ensure community participation in segregation of waste, the municipal authority shall organise awareness programmes through regular meetings of representatives of local residents, welfare associations and non-governmental organization.[49]

STORAGE OF MUNICIPAL SOLID WASTES

Generally solid waste collected is allowed to be accumulated in open leading to pollution and various unhygienic and insanitary conditions. The rule provides municipal authorities to establish and maintain storage facilities. The rule further lays down that storage facilities should take into account the quantity and nature of waste generated. Rules prohibit any sort of open dumping further storage facilities or 'bins' shall be 'easy to handle' design for handling, transfer and transportation. Bins for storage of bio-degradable waste shall be of green colour, for recyclable waste with white, for other waste shall be painted with black. Manual Handling of waste shall be prohibited, but if it is unavoidable, it shall be carried out under proper precaution and with due care for the safety of workers.

---

48 *Id*

49 *Id*, Schedule 2.

TRANSPORTATION OF MUNICIPAL SOLID WASTES

Vehicles for the use of waste shall be covered—not visible to public. It shall be prevented from scattering. Storage facilities—bins shall be cleaned on daily basis and should not be overflowed.

PROCESSING OF MUNICIPAL SOLID WASTE (BIODEGRADABLE)

'Suitable Technology or combination' shall be used by municipal authority to minimize burden or landfill. Bio-degradable wastes shall be processed by composting, vermicomposting, anaerobic digestion or any other appropriate biological processing for stabilization of wastes. Separate procedure has also been provided by Schedule IV of the Rules. Recoverable waste should follow the rule of recycling. Incineration, including pelletisation can also be used. Disposal of municipal solid wastes, land filling shall be restricted to non-biodegradable, inert waste and other waste that are not suitable either for recycling or for biological processing.[50]

Further special provisions are envisaged for landfills. Selection of landfill sites shall be based on examination of environmental issues. The landfill site shall be large enough to last for 20-25 years. The landfill site shall be planned and designed with proper documentation of a phased construction plan as well as a closure plan.[51] Landfill site shall be fenced or hedged and provided with proper gate to monitor incoming vehicles or other modes of transportation. The landfill site shall have wastes inspection facility to monitor wastes brought in for landfill, office facility for record keeping and shelter for keeping equipment and machinery including pollution monitoring equipments. Rules further provides for

50 *Id*

51 Schedule 3

specifications for land filling. Wastes subjected to land filling shall be compacted in thin layers using landfill compactors to achieve high density of the wastes.[52] In high rainfall areas where heavy compactors cannot be used alternative measures shall be adopted. Prior to the commencement of monsoon season, an intermediate cover of 40-65 cm thickness of soil shall be placed on the landfill with proper compaction and grading to prevent infiltration during monsoon. Proper drainage berms shall be constructed to divert run-off away from the active cell of the landfill. After completion of landfill, a final cover shall be designed to minimize infiltration and erosion. Same way elaborate provisions are envisaged for composting, incineration.[53]

The waste processing or disposal facilities shall include composting, incineration, pelletisation, energy recovery or any other facility based on state-of-the-art technology duly approved by the Central Pollution Control Board in order to control pollution. The incoming wastes at site shall be maintained prior to further processing. To the extent possible, the waste storage area should be covered. If, such storage is done in an open area, it shall be provided with impermeable base with facility for collection of leachate and surface water run-off into lined drains leading to a leachate treatment and disposal facility.[54]

Solid waste management has become a major environmental issue in India. In order to have a satisfactory, efficient, and a sustainable system of solid waste management, the following aspects need consideration,[55] Targeting waste reduction at source, Technological interventions, Social and

---

52 *Id*

53 *Id* Schedule 5

54 *Id*

55 Shaleen Singh, Sunil Pandey, Solid Waste Management in India : Status and Future Directions, TERI information Monitor on Environmental Science 6(1),(2000).

community efforts, Institutional and regulatory reforms and a Shift towards integrated waste management.

The classical approach to solid waste management (SWM) is a responsibility of municipal body which ignores several socio-economic aspects of most existing solid waste systems, thereby contributing to the problems of inefficiency and ineffectiveness of the system. The neo-concept of Integrated Sustainable Waste Management (ISWM) differs from the conventional approach towards waste management by seeking stakeholder participation, covering waste prevention and resource recovery, including interactions with other systems and promoting an integration of different habitat scales (city, neighborhood, household).[56] Essentially it implies that decisions about waste handling should take into account economic, environmental, social and institutional aspects not only in the waste production stage but also in its upstream and downstream stages. The integration can take place at various levels:

- The use of a range of different collection and treatment options,
- The involvement and participation of all the stakeholders, and
- The interactions between the waste system and other relevant systems such as industry.[57]

Suggestions for improvement should include aspects such as the quality of municipal services, systems of cost recovery

---

[56] Klundert A.V. and J. Anschutz, Integrated Sustainable Waste Management: The Selection of Appropriate Technologies and the Design of Sustainable Systems is not (only) a Technical Issue. Paper prepared for the CEDARE/IETC Inter-Regional Workshop on Technologies for Sustainable Waste Management, 13-15 July 1999 Alexandria, Egypt.

[57] Lardinios, I. and A. Klundert, Integrated Sustainable Waste Management. Paper for the Programme Policy Meeting, Urban Waste Expertise Programme.p 1-6, April 1997.

from users, and privatisation of collection systems[58] (Bernstein 1991; Schetenleib and Meyer 1992). Attention should also be given to technical aspects; upgrading of equipment used for waste collection, developing environr entally safer methods of disposal like composting, sanitary lanc.:.'l), while noting that large-scale solutions may not be feasible in areas with inadequate infrastructure.

SOCIAL AND COMMUNITY EFFORTS

In parallel to the formal system of waste management, there is an active informal network in Indian cities. This sector consists of waste pickers, Itinerant Waste Buyers (IWB's), waste dealers and wholesalers, and small recycling enterprises. The sector is driven primarily by market forces, and makes a significant contribution to the overall waste management process in Indian cities. Moreover, since the sector is labour intensive, it provides employment opportunities to a large group of people,[59] accounting for an estimated 1-2% of the workforce in large cities. Sudhir estimate the maximum number of waste pickers, IWBs and waste dealers in Madras at respectively 40,000, 3000 and 2700.[60] Although the role of the informal sector in waste collection is quite significant, the problem of SWM still lies in the partial collection of waste and inability of municipalities to handle the problem efficiently. [61]

---

[58] Bartone, C.R., J. Bernstein and F. Wright. "Investments in Solid Waste Management: Opportunities for Environmental Improvement." Policy, Research and External Affairs Working Paper No. 405, World Bank, Washington, DC, April 1990.

[59] Furedy, C. 1992. "Garbage: Exploring Non-conventional options in Asian cities." *Environment & Urbanisation* 4(2) (Oct.): 43-60, (1992).

[60] Sudhir, V., Srinivasan, G. and Muraleedharan, V.R. "Planning for Sustainable Solid Waste Management in Urban India." *System Dynamics Review* 10(3): 223-246,(1997).

[61] Ravindra, H. "Bangalore: An Environmental Profile." In K. C. Sivaram Krishnan (ed). Managing Urban Environment in India: Towards an Agenda for Action. Times Research Foundation, Calcutta,(1993).

The private sector is now becoming a key player in a number of industrialized nations. Private sector participation can help upgrade technical and managerial expertise, increase efficiency in operation and maintenance, and improve customer services, apart from bringing in the capital to support the government in its efforts at waste management. There is a strong case for comprehensively involving the private sector and encouraging it to invest in waste management in India. Private entrepreneurs in India are entering into activities like the collection and transportation of waste, and lately into treatment processes. Private companies can, along with door-to-door collection, take on such tasks as secondary collection and transportation including vehicle-maintenance.[62]

Non-governmental organizations can play an important role in effectively projecting the community's problems and highlighting its basic requirements for urban services. They could help in organizing the ragpickers into waste-management associations/groups under the supervision of the urban local body and the relevant residents' associations or market associations. There is a long history of community-based efforts in India to manage the urban environment. One successful NGO experiment to manage solid waste disposal is Exnora in Chennai. This started in 1989 when citizens, concerned about deteriorating environmental conditions, drew up an action plan to collect garbage. New containers were placed in the street and an awareness-raising campaign was organized. The rag-pickers, renamed city-beautifiers, were given loans by Exnora to purchase tricycles for door-to-door garbage collection and street cleaning. They received monthly salaries from the residents, from which they repaid the loans. Today, the city has 1,500 Exnora units, each

---

62 Shekdar. " Municipal Solid Waste Management - The Indian Perspective, *Journal of Indian Association for Environmental Management*, 26 (2): 100-108, (1999).

servicing 75,000 families or 450,000 people.[63] The strategies discussed above show the broad path that India needs to follow for efficient management of the solid waste, which is being increasingly generated in large quantities.

CONCLUSION

The explosion in world population is changing the nature of solid waste management from mainly a low priority, localized issue to an internationally pervasive social problem.[64] India currently is facing a municipal solid waste dilemma, for which all elements of the society are responsible. The community sensitization and public awareness is low. There is no system of segregation of organic, inorganic and recyclable wastes at household level. There is an adequate legal framework existing in the country to address MSWM. What is lacking is its implementation. In spite of a stringent legislation in place, open dumping is the most widespread form of waste disposal. The possible reasons for poor implementation could be a combination of social, technical, institutional and financial issues. Public awareness, political will and public participation as essential for the successful implementation of the legal provisions and to have an integrated approach towards sustainable management of municipal solid wastes in the country. Only then we can think of sustainable cities.

---

[63] Colon, M. and Fawcett, B., "Community-based Household Waste Management: Lessons Learnt from EXNORA's 'Zero Waste Management' Scheme in Two South Indian Cities", Habitat International, In press (Elsevier Publication).

[64] Kurian Joseph, Perspectives of Solid Waste Management In India,, International Symposium on the Technology and Management of the Treatment & Reuse of the Municipal Solid Waste, Shanghai, China, (2002).

6

CDM

Relevance to Sustainability and Energy Security in Developing Countries

INTRODUCTION

Climate change is an intimidating test due to several factors. It will have impacts on Health including human, terrestrial and aquatic. Further climate change is predicted to be unequally distributed. While certain regions could actually benefit from global warming by the expansion of arable lands and a decreased need for heating, tropical zones will suffer most from droughts, loss of water resources and the expansion of epidemics. Rising coastal levels due to the polar melting process will adversely affect the most low-lying developing countries like Bangladesh and small island states.[1] Nevertheless, whatever may be the impact they can only be predicted with huge uncertainties. Additionally to make matters worse Greenhouse gases are no local environmental pollutants and thus there is no domestic incentive to reduce their emissions. They arise in all sectors of an economy, which means that an efficient climate policy has to be cross- sectoral. Due to global mixing of greenhouse gases, an efficient climate policy must be done on a global scale.[2] Kyoto Protocol is first

---

[1] Henry Willis, Earths Future Climate. Oxford University Press, p 24.

[2] Sustainable Energy Briefing 8,White paper on the energy policy of South Africa, December (1998) available at http://www.earthlife.org.za/

Contd....

such major measure taken by global community initially adopted for use on 11 December 1997 in Kyoto, Japan and which entered into force on 16 February 2005. The Kyoto Protocol is an international agreement linked to the United Nations Framework Convention on Climate Change. The major feature of the Kyoto Protocol is that it sets binding targets for 37 industrialized countries and the European community for reducing greenhouse gas (GHG) emissions. These amount to an average of five per cent against 1990 levels over the five-year period 2008-2012. By setting such targets, emission reductions took on economic value. To help countries meet their emission targets, and to encourage the private sector and developing countries to contribute to emission reduction efforts, negotiators of the Protocol included three market-based mechanisms - Emissions Trading, The Clean Development Mechanism and Joint Implementation.[3]

The Clean Development Mechanism (CDM) of the Kyoto Protocol is the first global attempt to address a global environmental problem with a market-based mechanism. The CDM is a carbon credit market where sellers, located exclusively in developing countries, can generate and certify emissions reductions that can be sold to buyers located in developed countries. Since 2004 it has grown rapidly and is now a critical component of developed-country government and private-firm compliance strategies for the Kyoto Protocol. This chapter presents an overview of the development and current shape of the market with a special reference to India. This chapter addresses the issues of acceptability, workability and challenges surrounding the CDM especially by and in developing countries in relation to climate change. Climate change is more and more seen to be acting as an obstacle of

---

Contd. from previous page...

wordpress/wp-content/uploads/2009/02/se-8_white-paper-on-the-energy-policy-of-sa.pdf (last accessed on 4-10-2009).

[3] http://cdm.unfccc.int/about/index.html (Last accessed on 3-03-2009).

developing countries' development efforts, as the impacts of climate change threaten development process already incurred. Further they are in the process of growth and their rate of GHG emissions are expected to increase and are projected to exceed those of developed countries in few decades. At present only developed countries are mandated to reduce their GHG emissions. Even if they are able to achieve their mandate permanent stabilization of GHG concentrations and mitigation of climate change requires efforts and contribution of developing countries. However, now, as in the past, climate change is considered a long-term problem of little immediate importance to the governments of developing countries, for which the issues of poverty alleviation, security and economic growth are considered to be more urgent.

EVOLUTION OF CLIMATE CHANGE POLICY: IPCC PROJECTIONS TO KYOTO MARKET MECHANISMS

The international community struggles to address the discernable problem of global warming due to its natural and anthropogenic causes. Climate change emerged on the political agenda in the mid-1980s with the increasing scientific evidence of human interference in the global climate system and with growing public concern about the environment. In its first report in 1990, IPCC concluded that the growing accumulation of human made green house gases in the atmosphere would "enhance the green-house effect, resulting in an additional warming of the Earth's surface" by the next century, unless measures were adopted to limit emissions.[4] The United Nations General Assembly responded in 1990 by establishing the Intergovernmental Negotiating Committee for

[4] The Intergovernmental Panel on Climate Change (IPCC) has been established by WMO and UNEP to assess scientific, technical and socio-economic information relevant for the understanding of climate change, its potential impacts and options for adaptation and mitigation. It is currently finalizing its Fifth Assessment Report "Climate Change 2010". http://www.ipcc.ch/(last visited On 8-02-2007 at 10.20am).

Framework Convention on Climate Change (UNFCCC). It adopts as its goal the stabilization of GHG concentrations in the atmosphere at a level that will prevent dangerous anthropogenic interference with the climate system.[5] But it was rather a general approach and never specified emission targets or binding mechanisms and instruments of climate policy. The UNFCCC entered into force in 1994.[6] The convention established the Conference of Parties (COP) as its supreme body.[7] The UNFCCC required signing parties to create greenhouse gas (GHG) emission inventories and required a pledge from industrialized countries to reduce their emissions to 1990 emission levels by 2000. This pledge was not effective, and, in 1995, the parties to the UNFCCC adopted the Berlin Mandate, which declared their intent to agree by 1997

---

5 The Convention on Climate Change sets an overall framework for intergovernmental efforts to tackle the challenge posed by climate change. It recognizes that the climate system is a shared resource whose stability can be affected by industrial and other emissions of carbon dioxide and other greenhouse gases. The Convention enjoys near universal membership, with 189 countries having ratified. Under the Convention, governments: The Convention entered into force on 21 March 1994. http://unfccc.int/essential_background/convention/items/2627.php/(last visited On 18-02-2007 at 10.45am).

6 *Id* India signed the UNFCCC on 10 June 1992 and ratified it on 1 November 1993. Under the UNFCCC, developing countries such as India do not have binding GHG mitigation commitments in recognition of their small contribution to the greenhouse problem as well as low financial and technical capacities. The Ministry of Environment and Forests is the nodal agency for climate.

7 The Conference of the Parties (COP) is the "supreme body" of the Convention, that is, its highest decision-making authority. It is an association of all the countries that are Parties to the Convention. The COP is responsible for keeping international efforts to address climate change on track. It reviews the implementation of the Convention and examines the commitments of Parties in light of the Convention's objective, new scientific findings and experience gained in implementing climate change policies. http: // unfccc.int/essential_background /convention/convention_bodies/items/2629.php) (last visited on 15.02-2007 at 9.35pm).

on quantified emission limitations and reduction objectives for industrialized countries.[8] The Mandate deliberately excused developing countries from binding obligations on the ground that developed countries had historically contributed the most GHGs.[9] The Kyoto Protocol was adopted on December 11, 1997.[10] The Protocol entered into force on February 16, 2005 and targets six main greenhouse gases: carbon dioxide (CO_2), methane (CH_4), nitrous oxide (N_2O), hydrofluorocarbons (HFCs), perfluorocarbons (PFCs), and sulphur hexafluoride (SF_6).[11] This was the first meaningful step to address human contributions to global warming.

The Kyoto Protocol caps emissions of GHGs from thirty-seven industrialized countries (Annex I parties)[12] to return their emission of GHGs to an average of approximately 5.2% below their 1990 levels over the commitment period 2008-2012.[13] The Quantified Emissions Limitation and Reduction

---

[8] U.N. Framework Convention on Climate Change, Report of the Conference of the Parties on its First Session, held at Berlin from 28 March to 7 April 1995, Addendum, at 5, U.N. Doc. FCCC/CP/1995/7/Add.1 (June 6, 2007), available at http://unfccc.int/resource/docs/cop1/07a01.pdf.

[9] *Id*

[10] Kyoto Protocol to The United Nations Framework Convention On Climate Change, United Nations 1998, The Kyoto Protocol is an international agreement linked to the United Nations Framework Convention on Climate Change.

[11] *Id* Annex A Kyoto Protocol.

[12] *Id* Industrialized countries listed in Annex I are: Australia, Austria, Belarus, Belgium, Bulgaria, Canada, Czechoslovakia, Denmark, European Economic Community (Now European Union), Estonia, Finland, France, Germany, Greece, Hungary, Iceland, Ireland, Italy, Japan, Latvia, Lithuania*, Luxembourg, Netherlands.

[13] *Id* Article 3(1)The Parties included in Annex I shall, individually or jointly, ensure that their aggregate anthropogenic carbon dioxide equivalent emissions of the greenhouse gases listed in Annex A do not exceed their assigned amounts, calculated pursuant to their quantified emission limitation and reduction commitments inscribed in Annex B and in accordance with the provisions of this Article, with a view to reducing their overall emissions of such gases by at least 5 per cent below 1990 levels in the commitment period 2008 to 2012.

Objectives (QELROs), as the Protocol called them, range from 8 per cent for the European Union member countries collectively, 7 per cent for the United States, 6 per cent for Japan, stabilization for the Russian Federation and most countries with transitional economies, to an increase of 8 per cent for Australia and 10 per cent for Iceland [14] and allows those nations to meet their targets through bubble mechanism[15] emission trading,[16] joint projects,[17] and the Clean Development Mechanism (CDM).[18] The focus of this chapter is clean Development mechanism as it is the mechanism applicable for investment in developing countries, or countries not included in Annex I of FCCC. Currently a carbon market is developing rapidly as a step towards reducing and

---

14 *Id* Kyoto Protocol Annex B.

15 *Id* Article 4, Any Parties included in Annex I that have reached an agreement to fulfil their commitments under Article 3 jointly, shall be deemed to have met those commitments provided that their total combined aggregate anthropogenic carbon dioxide equivalent emissions of the greenhouse gases listed in Annex A do not exceed their assigned amounts calculated pursuant to their quantified emission limitation and reduction commitments inscribed in Annex B and in accordance with the provisions of Article 3. The respective emission level allocated to each of the Parties to the agreement shall be set out in that agreement.

16 *Id* Article 17, The Conference of the Parties shall define the relevant principles, modalities, rules and guidelines, in particular for verification, reporting and accountability for emissions trading. The Parties included in Annex B may participate in emissions trading for the purposes of fulfilling their commitments under Article 3. Any such trading shall be supplemental to domestic actions for the purpose of meeting quantified emission limitation and reduction commitments under that Article.

17 *Id* Article 6, For the purpose of meeting its commitments under Article 3, any Party included in Annex I may transfer to, or acquire from, any other such Party emission reduction units resulting from projects aimed at reducing anthropogenic emissions by sources or enhancing anthropogenic removals by sinks of greenhouse gases in any sector of the economy, provided that:a) Any such project has the approval of the Parties involved; (b) Any such project provides a reduction in emissions by sources, or an enhancement of removals by sinks, that is additional to any that would otherwise occur.

18 *Id* Kyoto Protocol Art 12.

stabilization of green house gases in the atmosphere to avoid dangerous global warming.

CDM has been considered as a surprise package of Kyoto, Among the various mechanisms prescribed under Kyoto for reduction of GHG, CDM is the only connection between the developed and developing countries under Kyoto protocol.[19] To a major extent the efficacy of Kyoto protocol and participation of developing countries in mitigation of climate change will depend on the functioning of CDM. Since emissions from developing countries will eventually surpass those from the Annex I countries, developing countries full participation in such a regime is crucial.[20]

ORIGIN OF CDM: CONNECTING DEVELOPED AND DEVELOPING NATIONS

CDM is completely the invention of a hurried last hour of conciliation at the COP3 in Kyoto in December 1997. As Farhana Yamin of the Foundation for International Environmental Law and Development notes, "no Party's proposals resemble the CDM." Prior to Kyoto, there were no published papers about it.[21] The CDM may perhaps too an extent be attributed to the proposal put forward by Brazil concerning the Clean Development Fund and various proposals concerning joint implementation. The Brazil had proposed the Clean Development Fund (CDF) in a meeting of the Ad Hoc Group on Berlin Mandate in 1997.[22] Under the

---

[19] Agus P. Sari and Stephen Meyers, Clean Development Mechanism: Perspectives from Developing Countries, *Lawrence Berkeley National Laboratory*, (1999).

[20] *Id*

[21] Yamin, F, "Issues and Options for Implementation of the Clean Development Mechanism". London, UK: Foundation for Environmental Law and Development: 3,(1998).

[22] Naoki Matsuo' "CDM In Kyoto negotiations", *Mitigation and Adaptation Strategies for Global Change*, Volume 8, Number 3 / September, (2003)

CDF proposal, Parties who fall short of complying with their recommended emissions commitments in the prescribed period are made to compensate by contributing to the CDF. The proceeds accumulated in the CDF are used by developing countries to foster sustainable development. A fraction of the proceeds, no more than 10 percent, would go to support adaptation measures in vulnerable countries.[23] At COP3 in Kyoto, CDF took the shape of financial mechanism for Annex I countries to gain emissions offsets to achieve their QELROs.

CDM has primarily two objectives: (1) Providing public or private entities from Annex I countries with flexibility in realizing their quantified emission limitation and reduction commitments and (2) Assisting non-Annex I countries (who host CDM projects) in achieving sustainable development.[24] Each CDM project activity is intended to result in real, measurable and long-term GHG emission reduction benefits that are additional to those that would occur in the absence of the project. The CDM is a market-based approach to the problem of global warming. It allows buyers, who may be Annex B parties[25] or firms within Annex B nations, to purchase credits from emission reduction projects carried out in non-Annex B nations. The CDM thus builds on experience derived from various regional markets for atmospheric pollutants, most notably the United States' experience with emissions trading under the Clean Air Act.[26]

As one of the flexible mechanisms of the Kyoto Protocol, the CDM allows for emission reductions in projects carried out in developing nations. The CDM was designed around the

---

[23] *Id*

[24] *Supra* note 10, Art 12 .

[25] Parties that have accepted targets for limiting or reducing emissions under the Kyoto Protocol are listed in Annex B to the Protocol. These Parties are known as Annex B Parties.

[26] The 1990 Clean Air Act is a piece of United States environmental policy relating to the reduction of smog and air pollution.

insight that the marginal cost of emissions reductions in developing, and especially rapidly developing, countries would be less than those faced by developed nations. It is designed to stimulate sustainable development in host countries while allowing industrialized countries to meet their reduction obligations abroad in a cost-effective way.[27] To that end, the developed country Party undertakes to provide resources to the developing country Party to enable project activities that contribute to sustainable development growth and reduction in GHGs. Credit for emission reductions generated under a CDM project only attaches if the reduction of GHG emissions is in addition to any that would occur in the absence of the project activity.[28] Those reductions are certified as credits, and the developing nation and investing company would determine how to share the credits.[29] Through this joint project, the developing country acquires a more efficient technology and capital investment in the plant. The company from the Annex I country acquires a share of CERs it can use to meet its emissions reduction commitments at home.[30] The CDM has gained momentum, and there are now more than 860 registered CDM projects in forty-nine developing countries.[31]

CDM-MODALITIES AND PARTICIPANTS

At the 7th Conference of the Parties to the UNFCCC (CoP-7) convened in Marrakech in November 2001, modalities, guidelines and procedures governing the use of flexible mechanisms, in particular, the CDM, were adopted as documented in the Marrakech Accords, with a view to a

---

[27] *Supra note10*, Art 12(2).

[28] *Id* Art 12 (5) (c).

[29] *Id* Art 12(5).

[30] *Id* Art 12(10).

[31] Alexander Adam, Technology Transfer To Combat Climate Change: Opportunities and Obligations Under TRIPS and Kyoto, 9 J. High Tech. L. 1 (2009).

prompt start to CDM project implementation.[32] The Marrakech Accords paved the way for Annex I Parties to ratify the Kyoto Protocol and thus bring it into force.

In the top hierarchical structure of the CDM is the CDM Executive Board (EB) set up under the aegis of the UNFCCC.[33] The CDM EB is completely accountable for an entire range of administrative functions with regard to the CDM. Operational Entities [34](OEs) will be assisting the CDM Executive Board in executing various functions that are pivotal to the smooth endorsement and function of CDM projects. The main functions of operational entities include validation of CDM projects in accordance by EB, verification of emissions reductions, certification of emissions reductions, etc.[35]

Host Country, where the project is proposed to be ultimately located has a vital role to play in the CDM progression. Firstly the ratification of the Kyoto protocol is a necessity for any entity to participate in the CDM. Further achievement of sustainable development is one of the major objective and rationale behind CDM. For this host country government has to provide a letter of endorsement stating that the proposed CDM projects is in line with sustainable development objectives of the country.[36] The criterion vary across countries, however, there are commonalities and they typically include: (1) social criterion such as improving quality of life, poverty alleviation and ensuring equity; (2) economic criterion such as contribution to local economic development and technology transfer; (3) environmental criterion such as

32 Modalities and Procedures for a Clean Development Mechanism as defined in Article 12 of the Kyoto Protocol, Marrakesh Declaration 2001.

33 *Id*

34 A Designated Operational Entity under the CDM is either a domestic legal entity or an international organization accredited and designated, on a provisional basis until confirmed by the COP/MOP, by the Executive Board (EB). http://cdm.un.fccc.int/DOE/index.html (Last accessed on 6-02-2009).

35 Supra note 34

36 *Id*

reducing fossil fuel use, energy conservation, improvement of local environment and health, and consistency with the national energy and environmental policy.[37]

In order to implement the mechanism each country is mandated to establish a National CDM Authority.[38] This highest authority will be in charge of setting up guidelines to ensure that CDM projects fulfill the basic objectives of GHG emissions mitigation and contribute to the sustainable development processes of the host country.[39] Lastly there exists facilitator's armed with the responsibility of making a CDM project happen through designing an appropriate project into a CDM project, identification of financing opportunities, assisting in the process of syndication between the two entities and the like.

CDM PROJECT CYCLE

CDM is a project-based system. It accomplishes its objectives of individual emission reduction projects that are each validated by designated third party verifiers and then registered by the mechanism's governing body, the CDM EB.

The CDM project cycle generally consists of eight major steps,

1. Design and formulation by project participants.
2. Approval by Designated National Authority (DNA).
3. Validation by first Designated Operational Entity (DOE A).
4. Registration by Executive Board (EB).
5. Preparation of financing by investors.

---

[37] Karen Holm Oslen "Clean Development Mechanisms Contribution to Sustainable Development, A Review Of Literature" available at http://www.cd4cdm.org/Publications/CDM & Sustain Develop_literature.pdf (Last accessed on 12-07-2008).

[38] *Supra* note 34 See Marrakesh.

[39] *Id*

6. Monitoring by project participants.
7. Verification/certification by second Designated Operational Entity (DOE B).
8. Issuance of Certified Emissions reductions credits.[40]

The first stage of a CDM project cycle is the identification of a potential project in a Non-Annex I country. Project proponents need to take into account any national or regional requirements for project eligibility and obtain approval from the host government. The next step for the project participant(s) to develop a Project Design Document, for which there is a specific template. This is a necessary condition to get a CDM project approved and registered by the EB, as eligible for crediting. The Project Design Document contains critical information about the project, including whether it has earned host country approval from the Designated National Authority. The Project Design Document also describes the project's baseline, including whether the project participant is using a standard methodology or proposing a new methodology, and sets forth the case for the project's additionality.[41]

The case is then verified by a third party agency, a DoE to ensure the project results in real, measurable, and long-term emission reductions.[42] With the validation and registration of

---

[40] Myung- Kyoon Lee, UNEP Riso Ctr. On Energy, Climate And Sustainable Dev., CDM Information And Guidebook 30 2d ed. 2004, available at http://www.cd4cdm.org/Publications/cdm%20guideline%C202nd%20edition.pdf (Last accessed on 6-09-2008).

[41] Kyle W. Danish, An Overview Of The International Regime Addressing Climate Change, Sustainable Dev. L. & Pol'y,10,(2007).

[42] *Supra* note 10, Article12(5)provides that Emission reductions resulting from each project activity shall be certified by operational entities to be designated by the Conference of the Parties serving as the meeting of the Parties to this Protocol, on the basis of: (a) Voluntary participation approved by each Party involved; (b) Real, measurable, and long-term benefits related to the mitigation of climate change; and (c) Reductions in emissions that are additional to any that would occur in the absence of the certified project activity.

CDM project, the project proponents, normally local developers and foreign investors, can start taking actions to implement the project in order to generate the Certified Emissions Rates (CER) benefits and other conventional financial income. Registered projects, and those that have entered the implementation phase, will be required to maintain internal monitoring systems to demonstrate they are achieving the emission reductions specified in the PDD. The monitoring report will be submitted to a different DOE (DOE-B) from the one who validated the PDD. The DOE-B will verify the monitored emission reductions and produce a verification report for the project. The EB will issue CERs to the project proponents within 15 days after the date of receipt of the verification and certification report from the DOE-B based on the difference between the baseline and the actual emissions.[43]

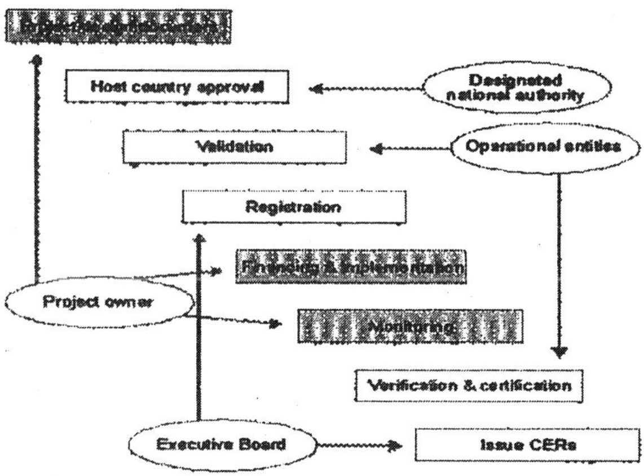

Figure 6-1: CDM Project Process

Source: Author

[43] *Id* Article 12(10), Certified emission reductions obtained during the period from the year 2000 up to the beginning of the first commitment period can be used to assist in achieving compliance in the first commitment period "certified emissions reductions" or CERs, are the output of CDM projects, and which under the terms of Article 12 can be banked from the year 2000, eight years before the first commitment period (2008-2012). http://unfccc.int/resource/docs/convkp/kpeng.html.

Each major step includes numerous sub-steps that result in an expensive and time-consuming process. These sub-steps increase transaction costs, which may deter CDM participation. Project design and formulation is one of the most time-consuming and expensive of these processes. At COP-7, it was decided that the following types of projects would qualify for fast-track approval procedures:

- Renewable energy projects with output capacity up to 15 MW.
- Energy efficiency improvement projects which reduce energy consumption on the supply and/or demand side by up to 15 GWh annually.
- Other project activities that both reduce emissions by sources and directly emit less than 15 kt CO_2 equivalent annually.[44]

Among the various reasons inhibiting CDM the CDM pipeline appears to be one of the major impediments. The CDM pipeline must be restructured in order to reduce the amount of time it takes for CDM projects to reach registration while also reducing the transaction costs associated with projects. Without such a restructuring, project participants may begin to withdraw their applications due to an inefficient and expensive process.[45]

In order to reduce pipeline during the nineteenth meeting of the EB in Bonn, Germany,[46] some proposed eliminating the EB or shifting some EB responsibilities to the World Bank Group or private sector entities that would bid to oversee CDM project activities. The alternative of shifting some responsibilities to private parties provides the following

---

[44] *Supra* note 34

[45] Mindy G. Nigoff, The Clean Development Mechanism: Does the Current Structure Facilitate Kyoto Protocol Compliance?, 18 Geo. Int'l Envtl.L. Rev,249, (2006).

[46] This meeting took place on May 11-13-2005. More information can be found online at U.N. Framework Convention on Climate Change, CDM: Meetings, http://cdm.unfccc.int/EB/Meetings (last visited Nov. 14, 2005).

possible benefits: some costs of review would shift to the private sector, increasing efficiency and allowing the EB to meet more frequently, transaction costs for project participants would decrease as a result of fewer costs for the EB, and fewer bureaucratic obstacles than currently associated with the CDM.[47]

Two approaches are proposed to relieve EB of some of its burden. The two different approaches are Compensated Reductions (CRs) and Environmental Capital Endowments (ECE). The Compensated Reductions approach provides post facto compensation in the form of carbon certificates to countries that elect to reduce national deforestation to below their 1980-1990 average.[48] In exchange for compensation, countries agree not to increase, or to reduce, deforestation in the future.[49] This approach provides emissions reductions and increased developing country participation without placing a burden on the EB for project registration. The Environmental Capital Endowment approach allocates "environmental capital" in the form of assigned amount units (AAUs) to developing countries that make early adoption of absolute emission caps below their BAU emissions target.[50] The ECEs could be used to finance investments in cleaner technology without participation in CDM projects.

CDM CURRENT STATUS: A CASE OF UNEVEN GROWTH

Since COP-7 was held in 2001, the Executive Board (EB) has been working as the CDM executive body to facilitate the CDM activities. Till early 2004, there were only 82 CDM

---

[47] *Id*

[48] Marcio Santilli et al., Tropical Deforestation and the Kyoto Protocol: A New Proposal, (2003), available at http:// www.amazonia.org.br/arquivos/ 92568.pdf.(last accessed on 1-02-2009).

[49] *Id*

[50] D. Dudek & J. Goffman, Envtl. Def. Fund. Emisssions Budgets: Building An Effective International Greenhouse Gas Control System (1997).

projects that reached the PDD development stage.[51] The first batch of CDM projects was registered in November 2004. After that, CDM has witnessed rapid progress. From a mere 82 projects the number of projects registered stands at 4733 projects (Table 6·0). The issued CERS stood at a whopping 285,544,051. Further there has been increase and diversification in number, type, and total volume of projects and emissions reductions in the CDM projects.

Table 6·0: Registered Projects and CERS as of May 2009[52]

Registered projects	1596
At Validation	2935
Process of registration	202
Issued CERs	285,544,051
Expected CERs (from registered projects until the end of 2012)	1,590,000,000

Among the Annex I countries the process of CDM was maximum utilized by United Kingdom accounting for a total of 29% of the Registered projects, Switzerland with a contribution of 20.8% came second (Fig. 6·2).

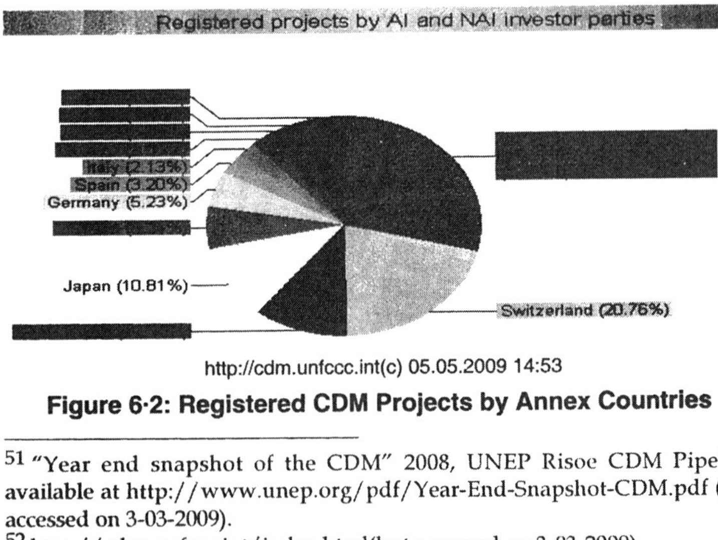

Figure 6·2: Registered CDM Projects by Annex Countries

[51] "Year end snapshot of the CDM" 2008, UNEP Risoe CDM Pipeline available at http://www.unep.org/pdf/Year-End-Snapshot-CDM.pdf (last accessed on 3-03-2009).

[52] http://cdm.unfccc.int/index.html(last accessed on 3-03-2009)

Regarding the Host countries China and India attracted the highest CDM projects. Together they accounted for 60 % of the total CDM projects registered. Next position went to Brazil and Mexico. Together these four countries claimed 74 % of the total projects. Critics point out this as a drawback of CDM. They believe that CDM does not provide equitable distribution and benefit to non Annex Countries. This mechanism merely benefits a handful of countries who are already on the developmental path (Fig. 6·3). It is to be noted that the representation of Africa is very negligible.

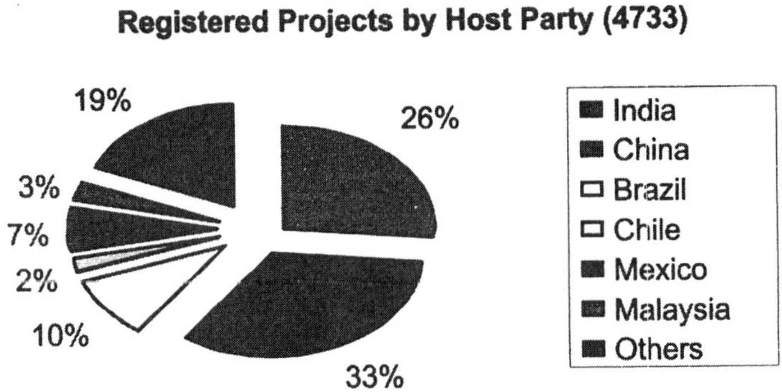

Figure 6·3: Registered CDM Projects by Host Party

Source: Author

The same anomaly and unequal distribution is depicted in the regional distribution of the CDM projects. 70% of the total registered project went to Asia Pacific region. (Fig. 6·4). This is hardly surprising as the major share of Project went to China and India. Second position was taken by Latin America which attracted 25%. For most regions, the dominance of one or two countries is clear: China and India dominate the Asian region; Brazil and Mexico have most projects in Latin America; South Africa has most projects in Africa.

Another pattern emerging from the project registrations that have occurred is the dominance of large projects in the CDM which is quite disturbing. Out of the 4733 project

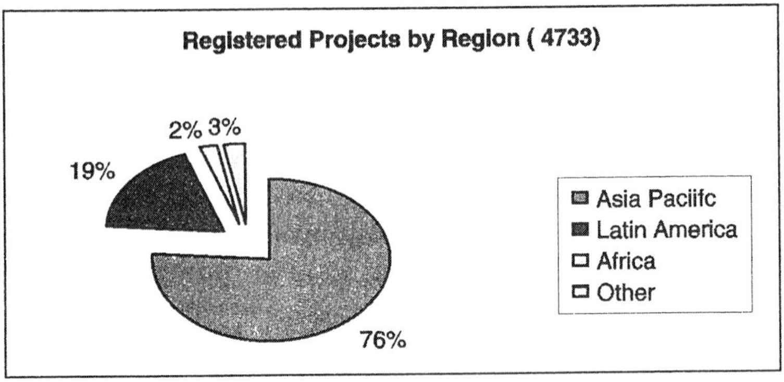

Figure 6·4: Registered Projects by Region

Source: Author

registered 56.94% was large scale projects. It is to be noted that compared to large scale projects, small scale projects are more economical, can be easily replicated in developing countries, and the time period spent on project formulation and implementation is less compared to large scale projects. Thus in order to promote sustainable development it is better to promote small scale projects (Fig. 6·5).

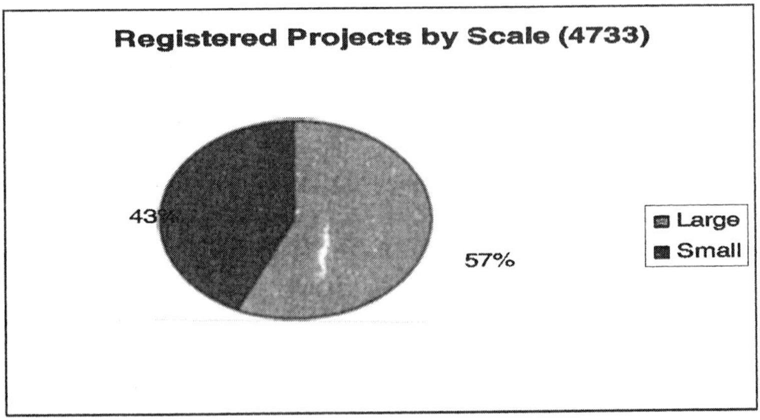

Figure 6·5: Registered Projects by Scale

Source: Author

While renewable energy projects do make up 60 per cent projects in the CDM project pipeline, they account for only 28 per cent of the emissions reductions produced. Among the renewable, the dominant CDM project technology is hydropower, which has a share in the November 2008 CDM pipeline of 26% (1098 projects), followed by biomass-based energy production (15%) and wind energy (14%). It is important to note that a significant proportion of the CERs generated by biomass power projects are from the CH_4 emissions that are avoided because biomass is burned rather than allowed to biodegrade. Much of the publicity surrounding the CDM has emphasized the number of renewable energy projects sponsored by the CDM while neglecting the relative volume of emissions, hence CERs produced and the relative scale of subsidy provided to various sectors. This emphasis provides a false picture of the true subsidy flows being generated by the international market for carbon (Fig. 6·6).

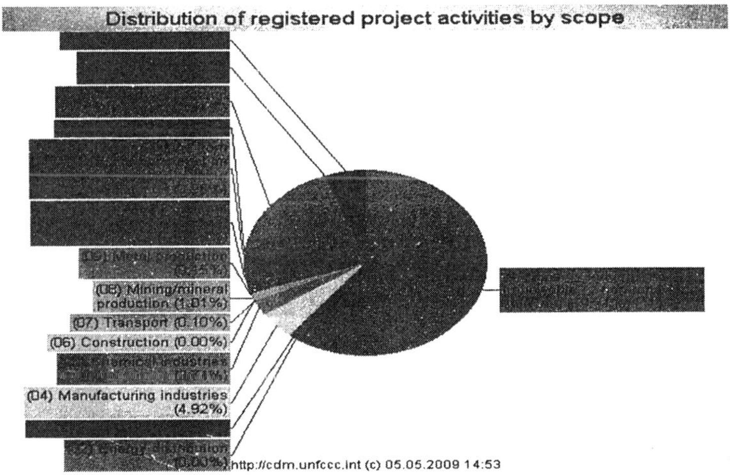

Figure 6·6: Distribution of Registered Projects

The original intent of the CDM was to spur development of low-carbon energy infrastructure in the developing world both through achievement of sustainable development goals and substitution for early retirement of expensive, high-carbon energy infrastructure in the developed world. [53] With respect to the quality of CDM projects, a number of problems were noted as a result of the early experience with the CDM. It comes as a surprise, then, to find then that the CDM pipeline bears only a partial relationship to this vision. Instead, the subsidy provided by purchase of CERs to date will largely ensure that high Global Warming Potential industrial gases such as trifluoromethane (HFC-23) and N_2O as well as CH_4 emitted by landfills and confined-animal-feeding operations (CAFOs) in non-Annex B nations are captured and destroyed. The very large projects dominating the supply of CERs are confined primarily to two relatively obscure industries--adipic acid and chlorodifluoromethane (HCFC-22) production. Adipic acid is the feedstock for the production of nylon-66 and releases abundant N_2O as a production byproduct.[54] If we take HFC-23, a by-product of the production of certain ozone depleting substances that is no longer be produced in developed countries, but are still produced in developing countries. These projects generally have a very large emission reduction potential since the global warming potential of hydrofluorocarbons is 11,700 times as large as for CO_2. Therefore, each tonne of HFC reduction delivers 11,700 CERs for trade under the CDM. It had become clear that the destruction of HFC-23 had become one of the more lucrative CDM projects, to a point that there was concern that it may provide an incentive for additional manufacturing of ozone

[53] R. A. Reimer, C. S. Slaten, M. Seapan, T. A. Koch, and V. G. Triner, *Adipic Acid Industry – N_2O. Abatement: Implementation of Technologies for Abatement of N_2O Emissions Associated with Adipic Acid Manufacture, in* NON-CO_2, Greenhouse Gases: Scientific Understanding, Control and Implementation, Proceedings of the Second International Symposium, Noordwijkerhout, Netherlands, 8-10 september (1999), 347, J. Van Ham, A.P.M. Baede, L.A. Meyer, and R Ybema Eds., 2000.

[54] A. McCulloch, *Incineration of HFC-23 Waste Streams for Abatement of Emissions from HCFC-22 Production: A Review of Scientific, Technical and Economic Aspects*, 2 (2005) *at* http://cdm.unfccc.int/methodologies/Background_240305.pdf (last visited May 4, 2006).

depleting substances in order to be able to generate CDM credits from the destruction of the HFC-23 by- product.[55] Another controversial type of project deals with carbon capture and storage. The Intergovernmental Panel on Climate Change (IPCC) completed a study of technical and methodological issues with respect to carbon capture and storage.[56] Specific issues raised include questions about permanence, and whether credits from carbon capture and storage constitute avoided emissions or a removal of greenhouse gases.

CDM: INDIAN SCENARIO

India acceded to the Kyoto Protocol on 26 August 2002.[57] India is currently enjoying the position of a major player in development of CDM projects. Vast technical human resource, strong industrial base, dynamic transparent & speedy processing by Indian DNA (NCDMA) for host country approval, various MOUs signed between India and various other Annex countries, wide spectrum of projects with different sizes-all offer advantageous position in terms of CDM potential. It is estimated that India can capture 10% of Global CDM market and Annual revenue estimated range from US$10 million to 330 million.[58] But there exists huge

[55] Meinhard Doelle, The Cat Came Back, or the Nine Lives of the Kyoto Protocol, 16 J. Env. L. & Prac. 261,July 2006.

[56] IPCC Special Report, Carbon Dioxide Capture and Storage, Summary for Policy Makers (Report by Working Group III) (Intergovernmental Panel on Climate Change, 2005).

[57] http://envfor.nic.in/cc/india_unfccc.htm, India is not required to reduce emission of Green House Gases under the Protocol under which basically the developed countries were required to reduce emissions of GHG by an average of 5.2 per cent below 1990 level by 2012. This Protocol was adopted in 1997 by the countries which are Parties to the United Nations Framework Convention on Climate Change (UNFCC). (last visited on 26-02-2009).

[58] Jitendra Kumar Singh, Clean Development Mechanism (CDM) And Carbon Trading In India, available at http://www.tce.co.in/Downloads/bro_pdf/papers/cdm_carbon_trading.pdf(Last accessed on 3-12-2008).

potential energy efficiency and conservation projects. For instance renewable energy generation can be from agriculture wastes, hydro and wind. Energy efficiency projects can also take various forms including: Increasing building efficiency through Green Building/LEED Rating, Increasing commercial/industrial energy efficiency through Renovation & Modernization of old power plants. It also consists of fuel switching from more carbon intensive fuels to less carbon intensive fuels and includes re-powering, upgrading instrumentation, controls, and/or equipment. Also thermal electricity generation and transmission offer countless opportunities to improve energy efficiency. The chemical industry and aluminium production will also allow reductions of industrial greenhouse gases with high warming potentials.[59] Transport sector comes next which includes Improving vehicle fuel efficiency by the introduction of new technologies - Changes in vehicles and/or fuel type, for example, switch to electric cars or fuel cell vehicles using CNG/Bio fuels. Switching the mode of transport also will qualify under CDM. For instance Metro in Delhi changes the mode of transportation to less carbon intensive.[60]

In pursuance of COP -7 decision Indian Government set up the National Clean development Mechanism. Its main function include,

- to invite officials and experts from Government, financial institutions, consultancy organizations, non-governmental organizations, civil society, legal profession, industry and commerce, as it may deem necessary for technical and professional inputs and may co-opt other members depending upon need,
- to interact with concerned authorities, institutions, individual stakeholders for matters relating to CDM,

---

59 *www.cdmindia.com/gtz/India%20CDM%20Potential.pdf* (last visited on 16-02-2009 at 9.10pm).

60 *Supra* note 60

- to take up any environmental issues pertaining to CDM or Sustainable Development Projects as may be referred to it by the Central Government, and
- to recommend guidelines to the Central Government for consideration of projects and principles to be followed for according host country approval.[61]

At present due to the effective operation of the Indian Designated National Authority[62] more than 297 project proposals with an emission reduction potential of over 297 million CO_2 have got official host country approval.[63] Out of this 83 % of the projects comes from the renewable energy sector which is clearly a positive signal (Fig. 6· 7)

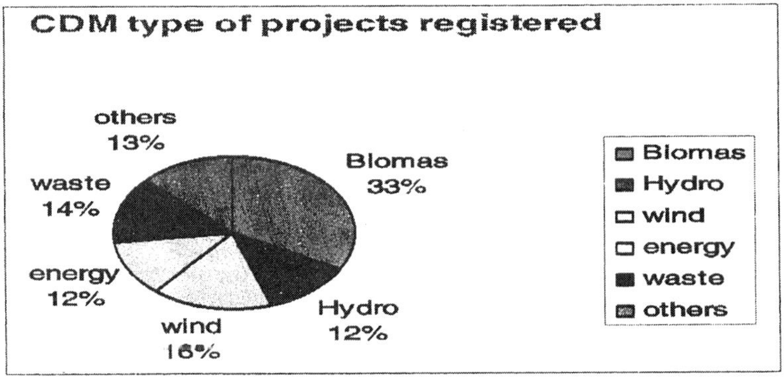

Figure 6·7 : CDM Type of Projects Registered in India

Source: IGES, 2008. "CDM Country Fact Sheet: India". Institute of Global Environmental Strategies

Two feasibility analyses were carried out with a view to promoting the CDM and analyzing its impact. The feasibility studies were undertaken in India by Community Forestry

[61] The Indian DNA is the National Clean Development Mechanism Authority available at http://cdmindia.nic.in/(Last accessed on 4-01-2009).

[62] *Id*

[63] http://www.cdmindia.com/India%20CDM%20Potential.pdf (last visited on 20-2-2009 at 4.50pm).

International, entitled *Communities and Climate Change: The Clean Development Mechanism and Village-based Forest Restoration in Central India* in Harda Forest Division in Madhya Pradesh and in Adilabad District, Andhra Pradesh. The studies reach slightly diverging conclusions concerning the implementation of CDM projects in India, but both promote the concept as a valid and viable income-generating activity for rural communities.[64]

An analysis of CDM mechanism and type of projects reveals that very likely India will be a major player. This is also true in the short and long run where India along with other developing countries may take on some commitments and where emissions trading may be an important component of the GHG abatement process. While there are several issues of concern from emissions trading, for India, on the whole may stand to benefit from participating in such trading. It is, however, important that we understand and address the problems in using a market-based approach to environmental issues in India on the whole, particularly with respect to monitoring and enforcement.

RESPONSE OF MARKET MECHANISM TO ACHIEVE SUSTAINABLE DEVELOPMENT

The Clean Development Mechanism (CDM) has been advocated as a means of reducing the costs of reduction GHGs and for supporting developing countries achieves sustainable development. It offers solution to certain environmental issues and act as a connecting mechanism between developed and developing world, in addition to benefiting investors and host country. At the same time the operation of CDM has generated considerable controversy.

---

[64] Emilo Caruso, Vijaya Baskar Reddy, The Clean Development Mechanism: Issues for Adivasi People in India, Forest Peoples Programme, (2005).

PANACEA FOR CLIMATE CHANGE

Many experts feel that with regard to climate change mitigation environmental concern should occupy the prime place, question of efficiency, justice and equity should be a late concern.[65] They point out various benefits for CDM. Firstly it is the only mechanism connecting developed and developing nations. Secondly if implemented properly it can help abate GHG emissions. Besides this developing countries can benefit from job opportunities, technology transfer, biodiversity and habitat protection or improvement of local air or water quality to achieve sustainable development. In addition investors may also be interested in promoting exports, expanding investment opportunities in developing countries, improving goodwill, and increasing standing in international negotiations.[66]

Exclusively from developing countries' perspective, CDM are supposed to bring in the following opportunities:

- Provide a new revenue stream for emission reduction projects by selling CERs, which can complement the conventional products and improve the financial viability of the projects;
- Promote the development, transfer and diffusion of environmentally-friendly technologies;
- Support project-based capacity building activities which can involve diverse stakeholders;
- Encourage the active participation of public sectors, private sectors and local communities, as well the cooperation with international entities;
- Gain learning experience and special knowledge through early CDM projects;

---

[65] Goldemberg, J (Ed)., *Issues and Options: The Clean Development Mechanism*, UNDP, New York, (1998).

[66] E.Parson, K.Fisher, Wanden, 1999, "Joint Implementation of Greenhouse Gas Abatement under the Kyoto Protocol's 'Clean Development Mechanism': Its Scope and Limits," *Policy Sciences* 32 (3), 207-224,(1999).

- Help define investment priorities in projects that meet sustainability goals;
- Contribute to public health and environment improvements; and
- Alleviate local poverty through income and employment generation.

Leaving aside these benefits there is a considerable dispute regarding the merits of CDM. Ensuring a continued acceptance of CDM by developing countries requires taking fully into account these concerns in the climate negotiations – or any international negotiation for that matter.[67] CDM cycle is a complicated CER "manufacturing process" with many participants involved in the process to generate and trade these real, measurable and additional assets. However, for the developing countries, due to project-based nature of the CDM and the capacity shortage reality, there are various institutional, legal and financial, environmental challenges to implement CDM.

PROJECT ISSUES: ADDITIONALITY

For qualifying under CDM there is a stricter requirement of additionality. It refers to "Reductions in emissions that are additional to any that would occur in the absence of the certified project activity."[68] The protocol lays down that only effective emission reductions must be validated, and for this purpose, real additional effort must be capable of being distinguished from spontaneous changes. Calculating additionality as proved to be one of the basic hurdles of CDM. Two major methodological approaches have been proposed in an effort to resolve this question. The first consists in

---

[67] Gupta, J., R.v.d. Worff, G. Junno, M. Hisschemoller, and P. Vellinga,, "International Policies to Address The Greenhouse Effect",. *Working Paper No.* 3. Amsterdam, the Netherlands: University of Amsterdam and Institute for Environmental Studies, Vrije University Amsterdam.,(1994).

[68] *Supra* note10 Article 12(5).

analyzing the projects on a case-by-case basis, while the second uses a more standardized process based on the definition of reference practices. Neither of these two approaches, however, appears wholly satisfactory, and the question of additionality still remains a major difficulty in terms of putting the CDM into practice; a major issue that should be resolved at the international level.[69]

Besides these, additionality concept is looked at from two angles as environmental and project additionality. These rivaling interpretations also add up to the controversy:

- In the first interpretation 'environmental additionality' means a project is additional if the emissions from the project are lower than the baseline. It generally looks at what would have happened without the project.
- In the second looks at it from the purview of project interpretation, meaning the project must not have happened without the CDM.

Many investors argue that the environmental additionality interpretation would make CDM simpler. But experts and NGOs have argued that this interpretation would open for free-riders and yield fewer credible emissions reductions. The current general interpretation is more in favour of Project additionality.This simply means whatever the projects and initiatives government undertakes as a matter of policy for the GHG reduction cannot qualify as CDM. For instance, if the Indian government specifies tough emission norms for buses, the public-transport sector does not get credits under CDM.[70] This current design provides countries with perverse incentives to keep polluting as long as they have the money to

---

[69] Denise Cavard, Pierre Cornut, Philippe Menanteau, "How Could Developing Countries Participate in Climate Change Prevention: the Clean Development Mechanism and Beyond", *Energy Review* (2001).

[70] Michaelowa, A. and K. Michaelowa.," Does climate policy promote development?" Climatic Change 84,(2007).

pick up carbon credits. In this respect, CDM actually works against sustainable development. This would effectively undermine the ultimate goal of the CDM.[71] It is necessary to establish a fair and reasonable objective method to judge whether the investments in the projects would not have happened anyway without CDM.

Determining a Baseline

Another important requirement of CDM is the calculation of baseline. The emissions reduction credited to CDM projects should be measured from hypothetical emissions that would have otherwise occurred without such projects known as the baseline of the project. The baseline may be estimated through reference to emissions from similar activities and technologies in the same country or other countries, or to actual emissions prior to project implementation.[72] Given the dynamic of the developing economies, establishing a credible baseline is a great challenge. Without a credible baseline the CER claimed also loses its credibility, which may threaten the CDM concept altogether. The partners involved in the project could have an interest in establishing a baseline with high emissions, which would yield a risk of awarding spurious credits.[73]

Fake CER issues

Each CDM credit - known as a certified emission reduction (CER) - supposedly represents one metric ton of carbon dioxide not emitted to the atmosphere. Governments can purchase credits directly or companies can buy them to comply with national-level legislation. Unfortunately, the CDM is failing miserably and is undermining the effectiveness

---

71 *Id*

72 Chomitz, "Baselines for Greenhouse Gas Reductions: Problems, Precedents, Solutions", Paper Prepared for Carbon Offsets Unit, World Bank, Washington, D.C., USA,(1998).

73 *Id*

of the Kyoto Protocol. Most "emission reduction" credits are "fake," from projects that do not actually reduce emissions. Billions of dollars have been transferred from taxpayers to undeserving project developers and a growing army of carbon brokers and consultants. In the process, the CDM is not only failing to support climate change mitigation and sustainable development in developing countries, but also provides industrialized countries with a way out of meeting their own domestic reduction obligations.[74]

MAXIMUM LIMIT

As indicated in Article 12, Kyoto Protocol, the carbon credits may be used by the Annex I Party to meet a *part* of the reduction targets.[75] This provision in effect discourages developed countries from framing domestic policies to reduce emission of GHGs. In order to further the objective of Kyoto Protocol, the level at which a cap or a limit would be levied on the CERs used by the Annex I Parties to meet their commitments has to be defined.

FINANCIAL ISSUES

In theory, the CDM allows for a drastic reduction of costs for the industrialised countries, while achieving the same amount of emission reductions as without the CDM. However, critics have long argued that emission reductions under the CDM may be fictive, and Transactions costs are higher than abatement costs. UNEP Studies point out that

---

[74] Schneider, L., "Is the CDM Fulfilling its Environmental and Sustainable Development Objectives? An Evaluation of the CDM and Options for Improvement." Report prepared for WWF by Öko-Institut, 5 November(2007).

[75] *Supra* note 9 Article 12(3) (b)Parties included in Annex I may use the certified emission reductions accruing from such project activities to contribute to compliance with part of their quantified emission limitation and reduction commitments under Article 3, as determined by the Conference of the Parties serving as the meeting of the Parties to this Protocol.

monitoring and verification costs could exceed the abatement cost savings [76] and in early 2007 the CDM came under fire for paying €4.6 billion for destruction of HFC gases while according to a study this would cost only €100 million if funded by development agencies.[77] Accountability and verifiability of emissions and credits etc is very difficult as stocks and flows of emissions are hard to quantify. Within the European Union, carbon is traded at US $26.7 per tonne of carbon dioxide equivalent. However, outside Europe CERs are being traded for a pittance. The only advertised price is by the World Bank, which pays about US$ 5.[78] To ensure whatever is done is *certified*, so that each CER is backed up with proof that it reduces an actual tonne of carbon dioxide equivalent. It is important for carbon credits to figure in the balance sheets of the developed world. So, the role of host governments can be minimized.

ENVIRONMENTAL CONCERNS OF GHG REDUCTION

The basic objective of CDM is the concern for environment in the form of achieving sustainable development and reducing GHGs. Critics argue that very purpose also get defeated in CDM projects. They argue that by shifting the problematic of climate change in this way, from its root causes to temporary external solutions, the real costs of climate change are not accounted for. Further rich countries avoid responsibilities at home through CDM and that it will actually increase emissions because the credits earned will allow rich countries to emit more, while developing countries are not tied to reduction at this stage.

[76] L.D.D Harvey and E.J.Bush, Joint Implementation an Effective Strategy for Combating Global Warming. Environment 38(8) 14-20.

[77] www.galtengroup.com/cdm.html(last accessed on 4-02-2009).

[78] *Ritu Gupta, Shams Kazi and Julian Cheatle, Acting pricey; Why is a tonne of tradable carbon worth US $5 in India and US $27 in Europe?, A side note to the main article, Newest Biggest Deal, Down To Earth Magazine, November 15, 2005.*

Article 12 of the Kyoto Protocol, lists achieving sustainable development as its basic objective. Assessing the sustainability impact of CDM projects using different criteria seems to be the major problem encircling all CDM projects. If one has to analyze the CDM rules for any guidance, it does not contain any definition of what sustainable development is. Countries hosting projects activities has been given the prerogative of unilaterally defining what they understand by sustainable development and whether projects assist them in achieving it or not. As the rules of the CDM state: it is the host countries' "prerogative to confirm whether a clean development mechanism project activity assists in achieving sustainable development".[79]

In all these years of implementation of Kyoto protocol, the basic criterion – CDM must "assist" developing countries for sustainable development – has got lost. Poor countries, with economic dependence, are forced into a senseless contest to assist the selling of credits, cheaply and as fast as possible.[80] The drive for 'cheap' credits and opportunity for GHG reduction is reflected in the kind of projects currently registered. Wind energy projects constitute 7 per cent of all projects; they bring less than 3 per cent of CERs being considered for sale. Over 50 per cent of CERs sold are to harvest fugitive gases – HFC-23 and N_2O – from industry. Another 26 per cent are for projects that harvest gases – largely methane – from landfills. There are less projects of afforestation. No high-end energy efficiency projects, and less of urban public transportation projects.[81]

---

[79] Markandya, A., Halsnaes, K. (Eds.) "Climate Change and Sustainable Development" Earthscan Publications, United Kingdom (2003) .

[80] The Clean Development Mechanism, A Users Guide Energy & Environment Group Bureau for Development Policy, UNDP 2003.

[81] Joy A. Kim, "Sustainable development and the Clean Development Mechanism: A South African Case Study". *Journal of Environment & Development,* Vol. 13, No. 3, p. 201-19,(2004).

CDM: THREAT ON SOVEREIGNTY

Besides various technical and financial concerns certain fundamental ethical issues act as an obstacle in the effective functioning of CDM. Most of these ethical issues are raised by developing countries that are young, many of which have obtained their independence from their colonizers in the 1950s. One of their prominent concerns was whether the CDM would diminish national sovereignty by allowing external forces to change the development paths of the host countries.[82] For the young developing countries Sovereignty is understood not just in legal terms but in terms of policies, control and principles. They fear that CDM may thwart their development process. This issue is particularly acute with regard to the land-use component of the CDM. They site various incidents from the past in support of their arguments and fear that domination by the big corporation will lead to another type of dependancy.

The notion of "national sovereignty" has come to the forefront of international arena in the 1972 World Conference on Environment and Development in Stockholm, and reaffirmed in the UNFCCC and Principle 2 of the Rio Declaration.[83] In this regard it is to be noted that CDM is a voluntary agreement.[84] The objection to CDM on the basis of

---

[82] *Supra* note 82

[83] In Agenda 21, "... States have, in accordance with the charter of the United Nations and the principles of international law, the sovereign right to exploit their own resources pursuant to their own environmental and developmental policies, and the responsibility to ensure that activities within their jurisdiction or control do not cause damage to the environment of other States or of areas beyond the limits of national jurisdiction." See UNCED [United Nations Conference on Environment and Development] (1992). Earth Summit: Agenda 21, The United Nations Programme of Action From Rio. Rio de Janeiro, Brazil: United Nations Conference on Environment and Development.

[84] Supra note 10 Article 12(5)Kyoto Protocol, provides Emission reductions resulting from each project activity shall be certified by operational entities to be designated by the Conference of the Parties serving as the meeting

Contd....

sovereignty arises from a distraught rather than any genuine basis. This distrust though not rational, nevertheless is a key feature of several arguments about CDM. In the climate change negotiations, distrust was enhanced by the US insistence for "meaningful participation from "key" developing countries. This call for developing country participation disregards the Climate Convention and the Berlin Mandate agreements that the US actively participated in shaping.

In addition it is also criticized that instead of relevant technology transfer to developing countries so that they are empowered to develop and produce themselves, the free-trade mechanisms will instead lead to further dependency. It is also feared that Developed countries will utilize this mechanism to dump outmoded and obsolete technology. Datas prove that approximately 36% of the registered and proposed CDM projects claim some technology transfer. But these projects account for about 59% of the annual emission reductions, so technology transfer is more common for larger projects.[85] The objection that CDM may distort host countries development priorities can be viewed as a repetition of the fear of interfering with sovereignty. It is feared that CDM investors would choose the most lucrative projects and if in the future the developing countries is obliged to reduce emissions it will be left with only high cost options.[86]

In summery we can deduce these ethical objections of developing countries are not based much on scientific basis but on their distrust of developed countries, This is hardly surprising as this is not confined merely to CDM but arises in any North -South discourse. Whatever may be the merit of

---

Contd. from previous page...

of the Parties to this Protocol, on the basis of: (a) Voluntary participation approved by each Party involved.

[85] Stephen Seres, Analysis of Technology Transfer in CDM Projects, prepared For UNFCC,(2008) .

[86] *Id*

these distrusts, the success of CDM will depend upon how far and fast we can ameliorate the doubts of Developing countries.

Post Kyoto Protocol : COP 9

SUGGESTIONS AND CONCLUSIONS

The main issue discussed in the chapter was whether the CDM is an effective market mechanism to assist an Annex I party, in meeting its Kyoto Protocol emission targets, while promoting sustainable development in non-Annex I countries.

The CDM in its current form is, from an environmental perspective, highly imperfect and is failing as a market because its rules, rather than producing real reductions, have accounting loopholes that allow participants to manufacture GHG credits at little or no cost beyond the payment of consultants necessary to surmount the necessary regulatory hurdles. Further, although it is supplying credits to developed signatories of the Protocol at prices less than they would otherwise be, the CDM is an excessive subsidy that represents a massive waste of developed-world resources. It is too late to change the structure of the CDM to address its shortcomings prior to the end of the first commitment period.[87]

Various suggestions can be made. A maximum collective cap should be fixed on the use of all the market options by the Annex I Parties for meeting their abatement commitments. This would guarantee that local effort is undertaken in the Annex I countries and they are not merely trying to escape responsibility by trading off emissions reductions from 'elsewhere'.

Since there exists no guidelines to determine the sustainability it is suggested, that at the international-level EB should provide a common minimum set of sustainable

---

[87] Gary C. Bryner, "Carbon Markets: Reducing Greenhouse Gas Emissions Through Emissions Trading", 17 Tul. Envtl. L.J. 267, 267 (2004).

development criteria that all CDM projects would have to satisfy. Further certain common eligibility criteria for climate change projects could be determined and agreed upon to ensure transparency and comparability across countries. At the same time, certain types of projects could be excluded due to the prevailing lack of international consensus regarding their desirability for instance, nuclear power projects and sinks. At the national level, the eligibility of potential CDM projects and sustainability is determined by host country governments. But it is important that the host country defines these national-level criteria according to its priorities, needs, and resource endowments. The contribution of the project to sustainable development could be determined by its performance in terms of access to technology, employment generation, income distribution, and provision of basic necessities like clean drinking water.

Certain other simplification process can be adopted to increase its efficiency. At the international level modify ER buyers' criteria to allow more projects into their pipeline and large institutional ER buyers with development agendas should increase operations in less developed countries. Further elucidating a post-2012 climate regime to allow for CDM projects with long time will also give a push to CDM.

At the national-level a simple, timely and transparent CDM project approval process will facilitate CDM process. It is essential to build national capacities for CDM project identification, design, approval, financing and implementation to reduce transaction costs.

Most developing countries are struggling to get out of poverty, which they view as caused partly by colonial exploitation. Alleviation of poverty and ensuring the availability of basic needs to their population are currently the main focus of their development process. This process is in no way to be compromised by foreign environmental interests. If CDM does not make its expected contribution to sustainable development, support for it is likely to erode. The wording of the Kyoto protocol seems to indicate that the chief beneficiary of the mechanism is the non-Annex I countries, which would benefit from accelerated development and also have an opportunity to participate 'meaningfully' towards the objectives of the Convention. It should be kept in mind however, that it is actually the Annex I Parties who would gain more by getting flexibility in meeting their commitments.[88] Thus, the CDM should not be considered the 'carrot' it is made out to be. The developing countries should put forward their perspectives in such a manner that they safeguard their interest and derive benefits in line with the objectives of CDM, while keeping in mind the ultimate aim of the FCCC.[89] Given the relatively poor performance, at least initially, of other markets for atmospheric pollution, the imperfect performance of the CDM is not entirely surprising and should not be a reason to abandon the system.

---

88 *Id*

89 *Id*

7

Green Consumerism in the Midst of Poverty

Eyewash on Climate Change Mitigation Strategy

INTRODUCTION

Situation 1: Anand Menon goes to purchase jeans, shopkeepers offered him an organic jeans giving out the details and benefit associated with it. They also made him aware of the positive impact this small action can have on climate change mitigation strategy. He thought for a while and decided to purchase it.

Situation 2: Harika goes to a beauty saloon for her routine monthly visit. Saloon offers her bio and green products. She consulted her friend who advised her to go for it as it is considered to be more beneficial.

These may be hypothetical cases but in our day to day life we often come across situations like this. "Go Green" has become the buzzword in any response to climate change encircling the economic social and political life. Stern report suggested that global warming is the biggest market failure we have ever seen.[1] There have been doubts whether market

---

[1] Stern, N., *Stern Review on The Economics of Climate Change. Executive Summary*. HM Treasury, London(2007) available at http://www.hm-treasury.gov.uk/sternreview_index.htm(last accessed on 12-01-2009).

by itself can solve this problem and market seems to have found a sound and just solution. From organic jeans and beauty products to green cars and nature friendly detergents, green consumerism is emerging as the hottest way to save the planet.

Sustainable development, and possibly a reversal of environmental degradation has become the buzzword encircling all political, social, and economical discourse in the last decade. The concept of Sustainability has rapidly taken a turn toward the cool and hip as more and more companies and firms roll out their lines of purportedly eco-conscious commodities resulting in "Green Consumerism" There is no consensus on the definition and meaning of Green Consumerism. But Green consumerism can be simply defined as "the conscious and deliberate choice to make certain consumption pattern taking into account economic consideration". People are realising that the pursuit of achieving sustainable and ethical levels of consumption is not only a prudent choice of utilising accessible reserves of raw materials, but also having substantial correlation within the "carbon sink" capacities of the global environment.

Green Consumerism can be looked at from two angles. Firstly, at producer's levels who can go green with using recycled materials or contributing money to environment reduction activities and launching green campaigns. Secondly, from the consumer's angle opting for environment friendly products. This chapter looks at the possibilities and limitations of Green consumerism especially from a developing world perspective where acute poverty still persists. Chapter will also analyse whether Green consumerism is only an eye wash that capitalism has to offer in the face of mounting ecological destruction or motivates manufacturers to be more environmental conscious in their production process.

GREEN CONSUMERISM: ORIGIN AND DEVELOPMENT

The concept of green consumption has been becoming fashionable and gaining popularity as a means of addressing and possibly reversing climate change. Generally speaking the term green consumerism describes people, products, or activities that are environmentally responsible. Green consumerism implicates that consumers attempt to minimize the negative environmental effects caused by the production, distribution, use, and/or disposal of the products they buy and producers aiming to manufacture environmental friendly goods to mitigate the depressing environmental consequences.[2] Though the concept of Green consumerism denotes both consumers as well as manufacturers emphasis is more on consumers since ultimately it is their consumption pattern that will determine the course of green consumerism. A green consumer is defined as "a person who, in his or her consumption behavior consciously attempts to have a neutral or positive effect on the earth, its environment and its inhabitants."[3] Green consumers consist of various categories: shoppers, retailers, customers, and shareholders, to name a few among them. Whatever may be the category they belong they all attempt to follow the three R's of reduce, reuse, and recycle.[4] Environmental or green consumerism can thus be defined as "individuals attempting to protect themselves and their world through the power of their purchasing and manufacturing decisions."[5] Basic idea behind the growing

[2] Grunert, K. G., & Thgersen, J. (Eds.). *Consumers, Policy and the Environment: A Tribute to Folke Olander*. New York: Springer,p234 (2005).

[3] Ronald Hill, Ainscough, Todd Shank Daryl Manullang, "Corporate Social Responsibility and Socially Responsible Investing: A Global Perspective," *Journal of Business Ethics*, Volume: 70 Issue: 2, p 165-174,(2007).

[4] Nicola Mendelson, "Using Strategic Alliances to Develop Credible Green Marketing", *Journal of Consumer Marketing*, Vol 12, p 4-8, (1995).

[5] Ottman, Jacquelyn, Meeting the Environmental Challenge, available at http://www.thefreelibrary.com/Green+consumerism:+the+trend+is+your+friend-a012865989(last visited on 2-03-2009).

popularity for Go Green is that when concerns of environmental problems penetrates intensely enough into the society, the demand and controlling power of the market will force producers' to green their products as well as production processes -- on fear of being redundant in the market-place by green-opting consumers. The assumption is that if this movement functions according to plan only those companies which adapt themselves to the critical demand for greenness will only sustain.

Green consumerism have been around for a long time, but the notion of a green consumer can, however, hardly be found in a standard textbook in macro economics. This is because environmental degradation is often characterized as an external effect which no single consumer can influence with his or her buying choices. The birth of the modern green consumer movement can be almost traced back to the time when Brundtland Report was creating increasing awareness of the global ecological crisis at the international level.[6] Green became the catch word that was used by marketers in the 1980 as a consequence of the increased awareness of the sustainability issues in politics, mass-media and financial groups.[7] This awareness and desire to act have gave rise to a phenomenon known as "green consumerism."[8] In fact in the late 1980's many manufacturing and retailing companies

---

[6] The Bruntland Report, or Our Common Future, is the report made by the World Commission on Environment and Development in 1987. It alerted the world to the urgency of making progress toward economic development that could be sustained without depleting natural resources or harming the environment. Published by an international group of politicians, civil servants and experts on the environment and development, the report provided a key statement on sustainable development, defining it as:'*[Development that meets the needs of the present without compromising the ability of future generations to meet their own needs]*.

[7] Ken Peattie, *Environmental Marketing Management : Meeting the Green Challenge*, Pitman Publishing, p. 24-40,(1995).

[8] Koenenn, To Market, to Market--With a Mission, L.A. Times, Mar. 15, 1990, at E1, col. 2.

experienced a dramatic change in consumers' tastes, in favor of products with less impact on the environment. Environmentally conscious consumers sought products that pose fewer threats to the environment. They wanted goods and packaging that used fewer resources and less energy to produce, whose production generated less pollution, and whose disposal did not contaminate the environment or aggravated the nation's landfill crisis.[9] By purchasing such products, they could feel that they are doing the right thing, usually without reducing the rate of consumption to which they are accustomed. During this wave of change firms realized that they could increase their profits by taking environmental concerns seriously. For instance in 1987 a British company called The Body Shop won the UK "Company of the Year" Business Enterprise Awards.[10] The Body Shop originates produces and sells naturally based skin and hair products as an outlet for "cruelty-free, minimally packaged, natural and bio-ingredient products.[11] The shop has grown rapidly since its inception in 1976 and in 1990 it had a turn over of $84.5 million and 457 outlets in the U.K and overseas.[12] The roaring success of Body shop was mainly due to its claim of using eco friendly process and products. The earlier concept that single consumers cannot affect the market underwent a dramatic change in1990s. Consumers have the power to change the market became the buoyant moto. Green consumerism was thus established as an essential ingredient of the corporate and business culture's responsible plan to save the planet. It came to be accepted as essential ingredient of the concept of corporate social responsibility. Further 1990's, there was an effort to define the contours of Green

---

[9] J. Elkington, J. Hailes, & J. Makower, The Green Consumer 5-11 (1990).

[10] Sandy Irvine, "Consuming Fashions? The Limits Of Green Consumerism", *The Ecologist*, Vol. 19, No.3, p. 88,(1989).

[11] Helen Peck, *Relationship Marketing : Strategy and Implementation*, Butterworth-Heinemann,p165, (1999).

[12] *Ibid*

consumerism and it was accepted to include words like:, "green markets", "green products" and "green consumers" [13] It is to be noted that the leniency in favour of Green was driven more by profit motive rather than by any genuine concern for environment. Thus unfortunately more than often the resulting products or services had more to do with marketing and less with the environment and were just an opportunistic way to respond to consumer needs.[14]

An analysis of literature brings forth that there are infinite definitions and ways of looking at the concept of Green Consumerism. Similarly there are varied meaning that can be assigned to the term green and what drives them. The important question is: Is it only environmental concerns that drives both consumers and manufacturers to opt for green products? Green consumers are driven by a combination of factors and various factors influencing the purchase and production of the green consumption decision it includes.

- Cultural factors: Consumption and production pattern get significantly affected by cultural factors. For instance if we take into account the consumption pattern of North Indians and South Indians remarkable differences can be noticed. It ranges from eating habits, clothes, selection of vehicles, construction of buildings etc. Specifically with regard to green consumerism, many people see it as a protest towards current customs and policies and practice of uncontrolled mass consumption. In Japan for instance there was a tradition of frugality, partly surprised by government's desire to stimulate consumption in the 80's only to reappear for environmental reasons in the 90's.[15]
- Social factors: According to leading economist Peattie, social factors play a close relationship with

[13] *Supra* note 7, p 26-30

[14] *Ibid*, p 26

[15] *Ibid*

consumption pattern. This comes from people's need for social acceptance and esteem. Peattie shows that the demand for fur coats might both increase and decrease due to the same need for approval from societal group.[16] But it is also to be noted that Social factors play a significant role generally with regard to the young generation of consumers. Cunningham supports this proposition. He characterized socially conscious consumers as "a pre-middle age adult of relatively high occupational status..., typically more cosmopolitan, but less dogmatic, less conservative, less status conscious, less alienated".[17]

- Environmental Concerns: Many times the green consumers do not conform to traditional market models which depict consumers as always acting in their best self interest. They might possibly look at the welfare and interest of public. Through their behavior they promote enhanced environmental performance from companies by buying specific green products. Further reports involving food poisoning, chemical residues, and unsafe cars have made environmental safety an important issue among consumers. This lead to an increase in consumption of products with specific advantages such as organic food or tamper-proof containers.
- Political factors: The policies of the government regarding protection and preservation of the environment often acts as a catalyst in encouraging a specific consumption pattern. The various policies, concessions and campaigns aimed at going green often encourages individuals to adopt a go green approach in their purchasing and production pattern.

---

[16] *Ibid*

[17] Anderson, W.T. Jr & Cunningham, W.H. "The Socially Conscious Consumer", *Journal of Marketing*, 36 (3) 23-31,(1972).

- Demographic factors: It includes education, age, knowledge, lifestyle etc. For instance young mothers are usually more careful towards the environment threats for the sake of their children. There is also a feedback loop from children to parents as many of the young generations get more environmentally oriented education in school. As for the lifestyle a study revealed that US consumers were "too busy" to care about the environment.[18] Further wealthy people esteem needs are particularly well expressed through green consumption as they stimulate peoples need for individuality and self respect.
- Convenience: Normally the cost of product to an extent determines the market of the product. High cost of green product often acts as one of the main barriers towards the green consumption. Peattie argues that if consumers are presented with more convenient green alternatives than the existing ones, can actually encourage green consumption even against the backdrop of high cost green products.

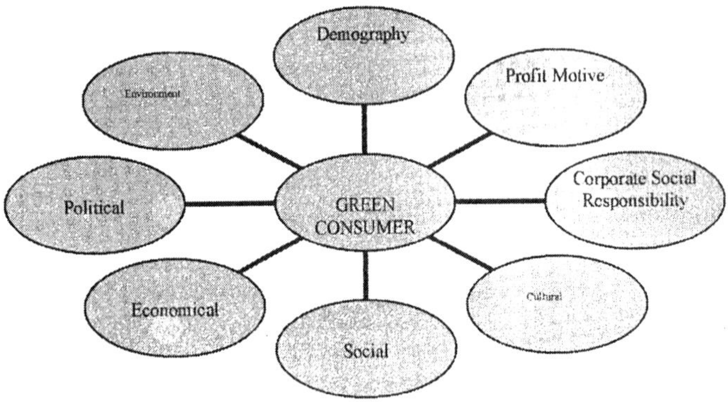

Figure 7·1: Factors of Green Consumerism

Source: Author

---

18 J.Seymour, H. Girardet, "*Blueprint for a Green Planet: How You can take Practical Action Today to Fight Pollution.*" Dorling Kindersley Ltd. London (1987).

From an investment and business perspective, Profitability assumes an important factor, Secondly there is a fair connection between ethical and environmental reasons leading to the green products and investment. Further the concept of corporate responsibility which is fact catching up also encourages green investment[19] (Fig. 7·1).

CURRENT STATUS OF GREEN CONSUMERISM

The Green Consumerism started in 1980 got real momentum in 1990's. In the opinion of some marketing commentators, green marketing has attained the status of a "revolution."[20] The growing environmental and sustainable concerns and awareness in political, economical and academics field has to an extent led to the emergence of the green investors and consumers, who are both environmentally aware and willing to invest their money on sustainable products and people willing and desiring to purchase green products. Various studies and data's point an upward trend of Green products both in terms of investment, manufacturing and consumption and perception of people.[21] Number of green products increased tremendously and constituted 5% of all new products introductions in the United States in 1989 and 10% in 1990.[22] Today consumers are bombarded with products claiming to be environmental friendly. At the same time consumers preference for green products has also seen steady rise. For instance there has been demand for organic produce doubled from 1988 to 1999 in the United States and business in recycled paper products and nontoxic

---

[19] Corporate Social Responsibility (CSR) refers to the responsibilities a business faces in relation to the social and environmental impact of their operations.

[20] The Green Marketing Revolution, ADVERTISING AGE, Jan. 29,(1991).

[21] Steven Yearley, *The Gwen Case: A Sociology of Environmental Issues, Arguments and Politics*, London, Harper Collins, p. 78-87, (1991).

[22] S.H Agarwal, *Green Management*, APH Publishing, p 109 (2005).

householdcleaners is brisk and growing.[23] Generally most of the attempts at capitalizing on the green movement have been focused on recycled products and environmentally friendly consumer products and packaging. Many companies have also looked to these as potential premium-priced goods.[24] On the positive side worldwide, buyers of energy-saving compact fluorescent lamps (CFLs) have eliminated the need for nearly 40 medium-sized, coal-fired power plants.[25] Software companies are going for a complete line up of eco - efficient consumer products same goes with the multi crore beauty industries.[26]

Any discussion on the status of green consumerism will not be complete without analysing the real consumer choices. Though there can not be any comprehensive study or data certain attempts have been made to ascertain the consumer preferences. It is to be noted that a 2007 Global Mackincey survey of 7751 people in Brazil, Germany, India and United States suggested that 87% of consumers are worried about the environmental and social impacts of the products they buy.[27] United States Environment Protection Agency estimated that Americans spend more than $115 billion on environmental friendly products.[28] Another study sponsored by green seal in 2009, regarding peoples perceptions and actual practice towards green consumerism revealed that out of the 1000 people surveyed half of them revealed that they buy same number of green products as they were earlier despite

---

[23] H. Patricia Hynes, 'The Pocketbook And The Pill: Reflections On Green Consumerism and Population Control", *Issues in Reproductive and Genetic Engineering,* Vol. 4, No. 1, p. 47-52, (1991).

[24] Mitchell Adrian, Kenneth Dupre," The Environmental Movement: A Status Report and Implications for Pricing", *SAM Advanced Management Journal,* Vol. 59,(1994).

[25] Peattie, Ken and Moria Ratnayaka "Responding to the Green Movement." Industrial Marketing Management, 21, 2, (May), 103--110. (1992).

[26] Is Green the way to go, *4Ps Business and Marketing,* p 73, Nov- Dec 2008.

[27] Supra note 7

[28] Supra note 22 See Agarwal, p110

economic slowdown.19% opined that they are in fact purchasing more.[29] The survey results were in fact surprising except energy from renewables in all other fields like minimal packing, green transportation, recycling, green personal care etc. People actually spent more than what they consider beneficial for the environment meaning their actual practice followed much more than their concern for the environment. While buying any product only nine percent of the surveyed took into account environment as a consideration, which does not depict a healthy trend. Majority of them went by the reputation of the product and brand loyalty. Tabular representation of the findings of the survey is presented below.

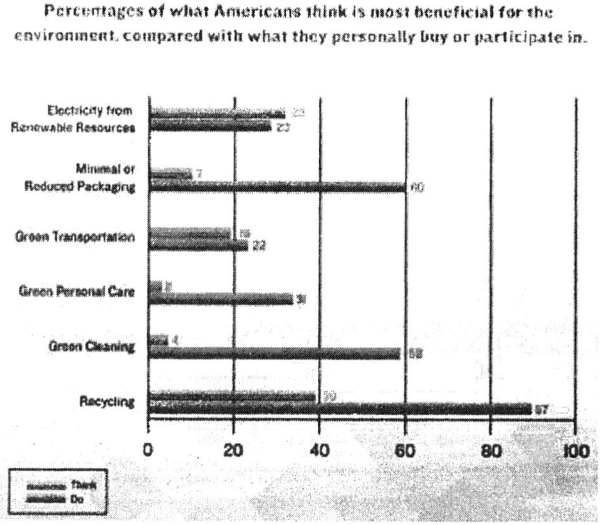

Figure 7·2: Perceptions of Americans Regarding Green Consumerism

Source: Enviro Media Social Marketing survey[30]

---

[29] Study commissioned by Green Seal and Enviro Media Social Marketing and conducted by Opinion Research Corporation reveals peoples' opinions and behaviors about products that claim to be environmentally friendly.

[30] National Green Buying research commissioned by Green Seal and Enviro Media Social Marketing. Conducted by Opinion Research Corporation January 23-26, 2009.

The above survey result pointed out that in USA consumers environmental concerns got converted into actions and they actually purchased green products. But as per the ICMR (Indian Council of Medical Research) survey conducted across five big Indian cities Delhi, Mumbai, Kolkatta, Chennai, Bangalore in 2008, there appears a mismatch between words and actions. Survey result reported that around 82% of the consumers are aware of green products but merely 32% of consumers surveyed actually brought or were willing to purchase those products that are marketed as green products.[31] Even when they bought it was mainly to reduce electricity bill or other like benefit. For instance people purchased Carrier air conditioner mainly to save electricity bill rather than due to any environment concern. Some of the major findings of the ICMR survey are presented below. More than 18% of the respondents were aware of more than 10 green products which is quite a high number. 34% of the respondents knew about 5 green products (Fig. 7·3) .

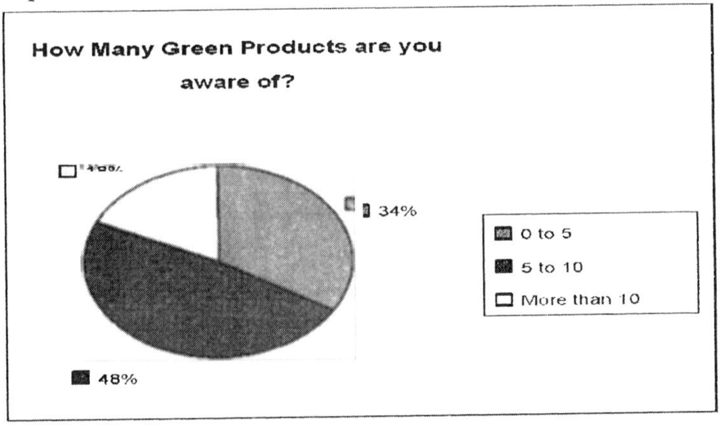

Figure 7·3: Green Products Awareness

Source: ICMR

About 57% of the respondents considered environment concerns as one of the primary consideration in affecting their

[31] *Supra* note 24

purchase.(Fig. 7·4). This figure can be considered to be a positive aspect for Indian consumerism since the awareness and concern of environmental issues can not be expected out of the majority of consumers in a developing country like ours. But the point for concern came from the fact that this concern did not actually converted into actions. Among the consumers who were interviewed only 32% have ever purchased a green products. (Fig. 7·5).

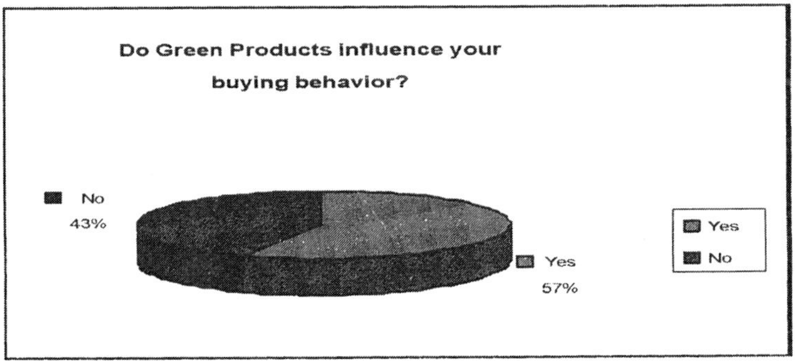

Figure 7·4: Influence of Green Products in Buying Behaviour

Source: ICMR

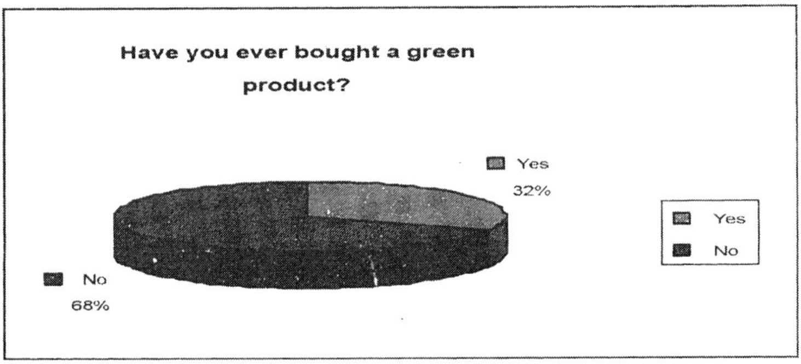

Figure 7·5: Green Product Purchase

Source : ICMR

The findings of the survey points that compared to USA the environmental awareness and concerns of consumers did not translated into actions in India. It may be because Indian consumers, majority of them belongs to the middle class category that has entered into the consumption arena only recently, are not in a position to afford normally premium priced environmentally friendly products.

Various studies carried out across the globe points to certain interesting features about green consumerism.

- Consumers across the globe have been asking for green products, meaning there has been a clear raise in demand for such products.
- Businesses have looked into the green process - generating corporate environmental profiles, monitoring and evaluating green performance, and improving corporate image as a result. The emerging concept of corporate responsibility encompasses being environmentally responsible as well.
- Green products have also increased competition among businesses to generate more environmentally friendly products.
- Marketing and demographic figures suggest that selling ethics pays and that this green marketing trend has room for expansion.[32] For instance the consumer electronics industry is finally discovering what other industries have known for a long time: "Green products are here to stay". Energy efficiency is a vital element for all products as they help in saving electricity and directly give benefits to the user.[33]

LEGAL FRAMEWORK FOR PROMOTION OF GREEN CONSUMERISM

Governments have also taken several measures that have

---

[32] Often referred to as "Caring Capitalism," companies tout their environmentalism. Patagonia switched in 1994 to organic cotton. Edward O. Welles *One Holdout: A White Knight Who Won't Quit*, Inc. Sept., p 45,(1998). Paul C. Judge, "It's Not Easy Being Green", *Bus. Week*, Nov. 24, p 180, (1997).

[33] *Supra* note 24, p 74

supported and facilitated such moves by businesses.[34] The United Nations recognizes that unsustainable consumption, along with the population growth, is one of the driving forces that threaten the sustainability of the planet. UNEP, for instance, created a Sustainable Consumption and Production (SCP) Branch solely for the purpose of increasing understanding about SCP and guide policy makers for appropriate actions.[35] The Living Planet Report 2006 which examines the health our ecosystem along with our present consumption pattern concludes that we need three planet Earths for us to sustainable our present wasteful way of life. Our Ecological Footprint goes way beyond the carrying capacity of our fragile planet.[36] At the international level ISO 14000 family addresses various aspects of environmental management including labeling, performance evaluation, life cycle analysis, communication and auditing. The Council on Economic Priorities annually publishes its "shopping for a better world" booklet with its findings on who's good and who's bad as a means of recommending the direction of consumer spending.[37] To increase consumer awareness, the

---

[34] Julia Hailes, "Green Consumer Guide", at the international symposium on Consumption and the Environment on 26 February, 1998 in Tokyo, Japan.

[35] UNEP's Sustainable Consumption programme was initiated in 1998 to supplement UNEP's production process oriented activities with activities on the demand side. Production and consumption policies were considered to be two sides of the same sustainability coin, which needed to be addressed in an integrated manner. The programme has led to the establishment of the Sustainable Consumption and Production Branch with its integrated life-cycle approach towards achieving more sustainable consumption and production patterns.

[36] Living Planer Report 2006,WWF.

[37] Council on Economic Priorities, Shopping for a Better World, (1992). The Council's rating key focuses on the following: giving to charity (top category is 2% or more); women's advancement (top category is 3 women as board members or officers); advancement of minorities (top category, is at least two minorities as members or officers); animal testing (top category is none); disclosure of information (top category is full disclosure on social programmes); community outreach (volunteerism in housing and education). South Africa (top category is no investment in South Africa);

Contd....

Government of India launched the eco-labelling scheme known as `Ecomark' in 1991 for easy identification of environment-friendly products.[38] Any product which is made, used or disposed of in a way that significantly reduces the harm it would otherwise cause the environment could be considered as Environment-Friendly Product.[39] The criteria follow a cradle-to-grave approach, i.e. from raw material extraction, to manufacturing, and to disposal. This scheme provides an incentive for manufacturers and importers to reduce adverse environmental impact of products. It also assists consumers to become environmentally responsible in their daily lives by providing information to take account of environmental factors in their purchase decisions.[40] In this scenario question also arises in case of deception or green washing followed by shoppers and manufacturers what remedy exists for the consumers. In India a consumer can normally proceed under torts for negligence. Further Indian penal code section 415 dealing with cheating also can be put to use.[41] But these provisions offer only an indirect protection to consumers. Specifically The consumer Protection Act

---

Contd. from previous page...

Environment (top category for recycling, alternative energy sources, green products and packaging); Family benefits (top category includes flexible time, benefits, child/dependent care assistance, education and seminars); Workplace issue (top category: good union relations and no union busting); Military contracts (top category: nothing over $500,000); Nuclear Power

[38] Ecomark Scheme of India available at http://envfor.nic.in/cpcb/ecomark /ecomark.html(Last visited on 3-02-2009).

[39] *Ibid*

[40] *Ibid*

[41] Section 415 Indian Penal Code 1860 provides that "Whoever, by deceiving any person, fraudulently or dishonestly induces the person so deceived to deliver any property to any person, or to consent that any person shall retain any property, or intentionally induces the person so deceived to do or omit to do anything which he would not do omit if he were not so deceived, and which act or omission causes or is likely to cause damage or harm to that person in body, mind, reputation or property, is said to "cheat". Explanation: A dishonest concealment of facts is deception within the meaning of this section.

1986(CPA) passed by the government of India offers some helps to the consumers in this regard. Section 2(r) of CPA prohibits unfair trade practices on the part of producers and manufacturers. Section provides that "unfair trade practice" means a trade practice which, for the purpose of promoting the sale, use or supply of any goods or for the provision of any service, adopts-any unfair method or unfair or deceptive practice including any of the following practices, namely the practice of making any statement, whether or all or in writing or by visible representation which, (i) falsely represents that the goods are of a particular standard, quality, quantity, grade, composition, style or model; (ii) falsely represents that the services are of a particular standard, quality or grade.[42] Any deception or green washing strategy concerning green claims can be easily brought under this. At the same time it is to be noted that India has not witnessed an increasing trend with regard to complaints involving green consumerism.

GREEN CONSUMERISM: AN ANSWER TO ECOLOGICAL CRISIS

Any discussion on green consumerism is revolved around one issue is it really going to be the answer to the impending ecological crisis. The answer is not easy to make. We need to brainstorm with various questions like for instance, does buying a Carrier Air conditioner, really help the planet? Do purchasing organic jeans really mitigate global warming?[43] Different people maintain different stand, some maintain that at best, it can have a limited impact in reducing environmental degradation and so postpone the ecological crisis. Others maintain that it could accelerate that crisis by creating new markets and thus increasing growth.

Supporters of Green consumerism argue that

---

[42] Section2(r) Consumer Protection Act 1986.

[43] Green Consumerism: Problem, Panacea or Just One of Many Good Ideas. Available at http://www.urth.tv/content/view/28333/215/ (last accessed on 10-02-2009).

Consumption of green goods often generates both private and public benefits. For example, a consumer benefits directly from consuming organic food because it is more nutritious and healthier with fewer side effects as it is free from pesticides and herbicide residues. Further eco friendly skin-care products like beauty creams and lotions are considered to be less harmful to the skin than traditional products containing chemical and synthetic ingredients.[44] Besides these direct private benefits to individual consumers, consumption and manufacturing of green goods also generates indirect public benefits as it helps in preserving the environment and mitigating the effects of climate change. On the production side, continuing with the example of organic food, organic farms are more sustainable and environmentally better than conventional farms because they do not release synthetic pesticides or herbicides into the environment.[45]

Critics maintain that Green consumerism is often the only thing capitalism has to offer in the face of mounting ecological destruction. Generally it is a hype generated by big corporate firms by using a few recycled materials or contributing money to an environment cause which are showcased as "concern for the environment and climate" Unfortunately, many companies pursue the "green" of increased revenue more eagerly than the "green" of a healthier environment. Such firms change their labels but little else. For them, the response to green consumerism "is not to market a new product or an innovation (the Japanese do that), but to be first with a [marketing] distinction."[46] They exaggerate or even fabricate the

---

[44] Jacobs, Michael, "The Green Economy: Environment, Sustainable Development and the Politics of the Future" Pluto Press, London, p 357, (1990).

[45] Parkash Chander and Subhashini Muthukrishnan, "Green Consumerism and Collective Action", National University of Singapore, Core Discussion papers,2007/58

[46] The Green Rush: Slowdown on Environmental Ties, ADVERTISING AGE, May 28, p 25 (1990)..

environmental qualities of their goods, letting their advertising rhetoric far outstrip their environmental contributions.[47] They also engage in "green-washing", in which companies lavishly fund on advertising campaigns to paint themselves "green" without altering their current polluting practices. This has been clearly proved in the survey conducted by Terra Choice Environmental Marketing Inc. In an effort to describe, understand, and quantify the growth of green washing, they conducted a survey of six category-leading big box stores.[48] They identified around 1,018 consumer products bearing 1,753 environmental claims. Of the 1,018 products examined, they found out that all but one made claims that are demonstrably false misleading. Based on the survey results, They identified six patterns in the green washing, and called it as the "Six Sins of Green Washing" which are as follows.

- Sin of the Hidden Trade-Off occurs when one environmental issue is emphasized at the expense of potentially more serious concerns.
- Sin of No Proof when environmental assertions are not backed up by evidence.
- Sin of Vagueness occurs when a marketing claim is so lacking in specifics as to be meaningless. Manufacturers often describe the environmental benefits of their products in the vaguest of terms: "environmentally responsible," "ozone friendly," "non-polluting," and even "lighter environmental formula."[49] Despite their ambiguity, these claims imply that they

[47] Garfield, Beware: Green Overkill, ADVERTISING AGE, Jan. 29, p 26,(1991).

[48] The "Six Sins of Green Washing", A Study of Environmental Claims in North American Consumer Markets, A Green Paper by Terra Choice Environmental Marketing Inc.

[49] Smith, Environmentalists, State Officers See Red as Firms Rush to Market "Green" Products, Wall St. J., March 13, p B1,Col3 (1990).

are based upon objective scientific investigation. Frequently, however, manufacturers have no reasonable factual bases for making vague, broad claims about a product's environmental "friendliness" or "responsibility," because such attributes are highly subjective.

- Sin of Irrelevance arises when an environmental issue unrelated to the product is emphasized.
- Sin of Fibbing when environmental claims are outright false.
- Sin of Lesser of Two Evils occurs when an environmental claim makes consumers feel 'green' about a product category that is itself lacking in environmental benefits.[50]

Among the various false green claims Sin of hidden trade off was practiced the most by firms and 57% of the green claims belonged to this category. Sin of no proof and vagueness came next constituting 26% and 11% respectively. Sin of Irrelevance and Fibbing occupied the last constituting 1% of the false green claims. (Fig. 7·6).

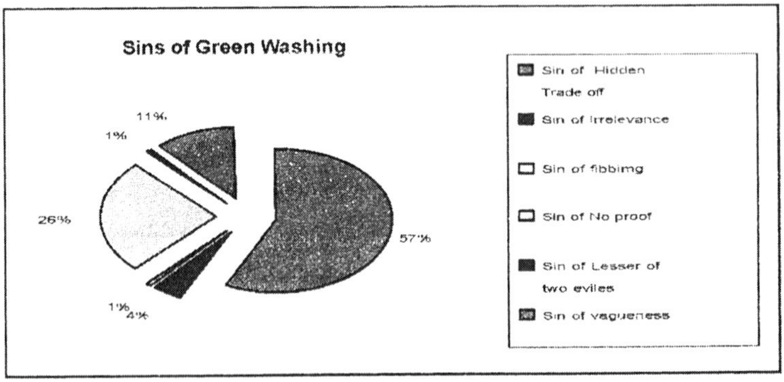

Figure 7·6: Sins of Green Washing

Source: Author data Green Paper by Terra Choice Environmental Marketing

50 *Supra* note 47

This will have the consequences of not only misguiding the genuine environment conscious consumers, but will thwart markets attempt to contribute in any manner for the climate change mitigation.

Besides these major corporate polluters who see a new marketing opportunity in green products are rapidly overtaking the green consumer movement. The steps in pollution prevention that some corporations are taking - waste reduction and minimization, energy efficiency, recyclable packaging - are necessary but long overdue. In fact they are being hailed and as an example of growing corporate responsibility for doing what they always should have done. Few, if any, are changing the products they manufacture to eliminate toxicity; where there are beneficial changes in product ingredient, they are driven more by green logo programmes than by industry initiative. Companies strive to brand their corporate image and products with a green seal of approval that speaks more to political correctness than to any meaningful measure of corporate responsibility.[51] Other problems they point out include some rare instance firms have resorted to Deliberate Lies. One US-based hamburger chain claimed that all its beef was 'US beef' thus countering charges about rainforest being cleared in Central and South America for beef cattle grazing. This claim apparently exploited a legal technicality, whereby beef imported into the USA and then processed can be classified as 'US beef'.[52] Manufacturers of disposable nappies have claimed that their products are environmentally-friendly because they have begun to use pulp made using a chlorine-free, dioxin-free process. This claim overlooked other factors: the chlorine-free, dioxin-free process, intoxicity tests, killed more fish, through the release of waste water from factories, than the other process. As well, the

[51] Jon Entine, *Corporate Ethics and Accountability*, 5 At Work, p 3, (1996).

[52] Teaching for a sustainable world :International edition, Module 16, available at http://www.environment.gov.au/education/publications/tsw/modules/module16.html(last accessed on 16-03-2009).

potential problem of enormous quantities of human faeces being buried in landfill dumps was overlooked.[53]

The contribution of green products in market is substantially less. For instance organic agriculture in the United States supplies only 4% of commercial food.[54] Consumer concerns about environmental issues may not translate into pro-environmental behaviours. This has been proved by ICMR survey. However, those individuals with a strong belief that their behaviour will result in a positive outcome are likely to engage in such behaviours in support of their concerns for their environment.[55] However, despite increased environmental consciousness, the demand for green products has been shown to be uneven across different market segments.[56] This came up in a Finnish Climate Change Communication campaign which reported the attitude change of Finnish people: people's readiness to make changes in their lifestyles to a more green direction has increased while in the same time there have not been any real changes in their consumption habits.[57] This demonstrates the fact that attitudinal and behavioural changes are different. The key is to convert attitudinal change into action.[58]

Besides these from a third world perspective green consumerism brings more issues. Shifting from non-green to

---

[53] *Ibid*

[54] H. Patricia Haynes, The Race to save the Planet: Will Women lose, Women's Studies International Forum,Volume 14,Issue 5, p 473-478,(1991).

[55] Ana Laskovaa, Perceived Consumer Effectiveness and Environmental Concerns, Proceedings of the 13th Asia Pacific Management Conference, Melbourne, Australia, 2007, 206-209.

[56] Ottman, Jacquelyn A "Industry's Response to Green Consumerism." Journal of Business Strategy, 13, 4, (July/August), 3--7 (1992).

[57] The Finnish Climate Change Communications Programme 2002-2007 has been an important part of Finland's national climate strategy. The programme has aimed to increase awareness of climate change, its impacts, and ways to mitigate them. It was carried out by Taloustutkimus Oy, commissioned by the Ministry of Trade and industry.

[58] *Ibid*

green behaviour comes at a premium costs to the consumer. These extra costs thrust upon the consumer are not just in terms of increased monetary consideration but in terms of shifting a particular brand, spending valuable time in learning etc. Green consuming requires sacrifices in terms of benefit like reducing the consumption of good and accepts lesser quality and durable goods in favour of quality products. For example, baking soda may not leave your bathroom smelling clean as a daisy, but it is a non-toxic, alternative cleaning agent. In a country like India where poverty still persists in large numbers any discourse on green consumerism will be limited to only a small section. In countries where people can barely afford the necessary items asking them to opt for green products and paying premium and compromising on quality many times seems impossible.

Further Green consumerism, by its very nature, cannot challenge the profit based capitalism and contribute to environment. Every enterprise under capitalism is guided by profit motives and green companies are no exception. If they have to survive they must make profits. At the same time this is not to suggest that green companies are by nature incapable of making any profits. For instance Body shop made substantial on the basis of green drive. Sustainable development has to be managed by the government and corporates alike and their effort should be complementary to each other. Sustainable Development has to be embedded into the corporate DNA companies. Corporates need to be transformational and need to act instead of only paying lip service. But generally until the market based capitalist society can think beyond profit solutions like green consumerism will be about as effective as fighting a forest fire with a water pistol. Such solutions are doomed to failure because they promote individual responses to social problems, problems that by their very nature require collective action, and deal only with the symptoms, rather than focusing on the cause of the problem in the first place.[59]

"Green marketing" and corporate environmentalism may conspicuously aim at altering consumption patterns but will never encourage any overall reduction of consumer spending.[60] Buying with environmental consciousness should be, first, buying less, and then, buying differently. Green companies, consultants, and catalogues do not prompt us to ask ourselves if we truly need something or simply want it.Once "green consumerism" is enshrined as the means to our ecological salvation, producers of less than green products can simply "throw up their hands and say 'the consumers made me do it'".[61] Achieving a sustainable future is not a task that can be left to the markets; it requires a renewal of non-material values, and political action by people as citizens, not simply as consumers.[62]

Andrew Watson one of the critique of green consumerism sums up very eloquently as follows:

"green consumerism, which is largely a cynical attempt to maintain profit margins, does not challenge capital's eco-cidal accumulation, but actually facilitates it by opening a new market. All products, no matter how 'green', cause some pollution, use some resources and energy, and cause some ecological disturbance. This

---

[59] E.7 Can green consumerism stop the ecological Crisis?, available at http://flag.blackened.net/intanark/faq/secE7.html(last accessed on 3-03-2009).

[60] Sharon Beder, *Global Spin: The Corporate Assault on Environmentalism,* Devon: Green Books, p 176,(1997).

[61] Timothy W. Luke, "Green Consumerism: Ecology and the ruse of Recycling." In the Nature of Things: Language Politics and the Environment, eds. Jane Bennett and William Chaloupka. Minneapolis: University of Minnesota Press, 154-172,(1993).

[62] Michael Hannis, Myth of Green Consumerism, Consumption, Community and Free markets, available at http://www.lancs.ac.uk/depts/philosophy/awaymave/onlineresources/myth%20of%20green%20consumerism%20(m%20hannis).rtf (last visited on 2-03-2009).

would not matter in a society in which production was rationally planned, but in an exponentially expanding economy, production, however 'green', would eventually destroy the Earth's environment. Ozone-friendly aerosols, for example, still use other harmful chemicals; create pollution in their manufacture, use and disposal; and use large amounts of resources and energy. Of course, up to now, the green pretensions of most companies have been exposed largely as presenting an acceptably green image, with little or no substance. The market is presented as the saviour of the environment. Environmental concern is commodified and transformed into ideological support for capitalism. Instead of raising awareness of the causes of the ecological crisis, green consumerism mystifies them. The solution is presented as an individual act rather than as the collective action of individuals struggling for social change. The corporations laugh all the way to the bank." [63]

CONCLUSION

The mass enthusiasm for green consumerism is gaining. But the market is still dominated by the majority who tend to resist green marketing and a minority who tend to respond to it. If this is the case then there is little prospect for green consumerism succeeding as a method for solving the environmental crisis. The idea of green consumerism envisaged a mass fundamental attitudinal change in consumption and production and called for the mass consumer market to demand a complete restructuring of manufacturing along environmentally sound lines. A handful of consumers cannot take the green consumerism to its successful and logical culmination. A collective action is needed even to slow down the ecological crisis. Further in the absence of economically viable green products green consumerism will remain as the monopoly of rich which will further complicate the issue. This issue assumes importance

---

[63] *Id*

from the perspective of third world countries where poverty and inequality still persists. For instance taking the case of India who will be in a position to demand and consumer green products, a rickshaw puller who struggles to meet both ends meets definitely not. Green products can only be afforded by the rich and certain sections of growing middle class of India.

There are a number of points that can be drawn. The first is that green consumerism is still in its infancy specially in developing countries. Any initiative on green consumerism requires taking account of social, cultural and economic factors of society. Each country has its own traditions, norms and taboos. The programmes and studies on green consumerism should examine producers and consumers' attitudes and values towards environmental issues in the cultural barriers in the target country must be identified. Even after this it is not certain how far green consumerism solve the ecological crisis. Green consumerism by itself cannot be an answer to the ecological crisis. At present green consumerism is leading to a new economy and not a clean economy. Efforts to develop green products deserve definite support and praise. But in the context of the global economy, sustainable and green consumer actions alone will not be enough to conserve the planet and mitigate the effects of climate change. A bold new approach, which is based on a reductionist approach towards consumption and conservation strategies, is needed.[64] The more we consume, the more we affect the environment and the less "green" we become. The greenest path, therefore, is to consume less, not simply to consume the same amount differently, till that happens Green Consumerism will only offer lip service to save planet.

It is too easy to be cynical about green consumerism because of little shortcomings. Nevertheless; we should resist

[64] Jared Hardner and Richard Rice, "Rethinking Green Consumerism", *Scientific American*, May (2002).

this cynicism and preserve the green consumer movement. No matter how much we reduce our consumption, we always need to consume. When we do, we can make choices that entail different effects, both in type and degree, on the environment we want to protect. Green consumerism holds the promise of reducing these effects market wide by allowing consumers to demand products that pose less environmental harm than others. We must therefore prevent little green lies from derailing this movement. Only by inhibiting the ability of manufacturers to stretch the truth can we restore a degree of faith in a movement presently threatened by misinformation, disillusionment, and cynicism. Only when our choices are informed accurately and truthfully can we hope to become a society of "greener" consumers. For this policy makers and vigilant civil society has a major role to play.

As of now the ideas of green consumerism can be said to be subliminally contained in a recent press advertisement for Master Card. A picture of the earth-in-space is prominently featured with consumer goods like running shoes and a mobile phone melted onto the continents, and completely covering them. The headline reads "Master Card introduces Master values. A whole world of savings." The message seems to be `Junk the planet; we've got some serious shopping to do.'[65]

[65] Richard Gosden, "Greening all the way to the Bank", *Arena Magazine*,No 16, p 35,(1995).

Index